Intelligent Quantum Information Processing

The book discusses the foundations of intelligent quantum information processing applied to several real-life engineering problems, including intelligent quantum systems, intelligent quantum communication, intelligent process optimization, and intelligent quantum distributed networks.

This book:

- Showcases a detailed overview of different quantum machine learning algorithmic frameworks.
- Presents real-life case studies and applications.
- Provides an in-depth analysis of quantum mechanical principles.
- Provides a step-by-step guide in the build-up of quantum inspired/ quantum intelligent information processing systems.
- Provides a video demonstration on each chapter for better understanding.

It will serve as an ideal reference text for graduate students and academic researchers in fields such as electrical engineering, electronics and communication engineering, computer engineering, and information technology.

QUANTUM MACHINE INTELLIGENCE SERIES
SERIES EDITORS

Siddhartha Bhattacharyya,
VSB Technical University of Ostrava, Czech Republic and Algebra
University College, Zagreb, Croatia

Elizabeth C. Behrman,
Wichita State University, USA

Quantum Machine Intelligence
Siddhartha Bhattacharyya, Mario Koppen,
Elizabeth C. Behrman, Iván Cruz-Aceves

Hybrid Computational Intelligent Systems
Modeling, Simulation and Optimization
Siddhartha Bhattacharyya

Intelligent Quantum Information Processing
Siddhartha Bhattacharyya, Iván Cruz-Aceves, Arpan Deyasi,
Pampa Debnath, Rajarshi Mahapatra

For more information about this series, please visit: www.routledge.com/
Quantum-Machine-Intelligence/book-series/QMI

Intelligent Quantum Information Processing

Edited by
Siddhartha Bhattacharyya,
Iván Cruz-Aceves, Arpan Deyasi,
Pampa Debnath, and Rajarshi Mahapatra

CRC Press
Taylor & Francis Group
Boca Raton London New York

CRC Press is an imprint of the
Taylor & Francis Group, an **informa** business

First edition published 2024
by CRC Press
2385 NW Executive Center Drive, Suite 320, Boca Raton FL 33431

and by CRC Press
4 Park Square, Milton Park, Abingdon, Oxon, OX14 4RN

CRC Press is an imprint of Taylor & Francis Group, LLC

ISBN: 978-1-032-39267-7 (hbk)
ISBN: 978-1-032-44632-5 (pbk)
ISBN: 978-1-003-37311-7 (ebk)

DOI: 10.1201/9781003373117

Typeset in Sabon
by Apex CoVantage, LLC

Siddhartha Bhattacharyya would like to dedicate this volume to his loving wife, Rashni.

Iván Cruz-Aceves would like to dedicate this volume to his children, Ivan and Yusef, hoping that they keep their energy and curiosity!!!

Arpan Deyasi would like to dedicate this book to his guru, Swami Atmapriyananda Maharaj, who first introduced him to the world of Quantum Physics.

Pampa Debnath would like to dedicate this book to her respected father, Mr. Parimal Debnath; her mother, Mrs. Krishna Debnath; her beloved husband, Mr. Snehasis Roy; and her kid, Master Deeptanshu Roy.

Rajarshi Mahapatra would like to dedicate this volume to his wife, Sarmila.

Contents

9 Quantum Microwave Engineering: A New Application
   Area of Quantum Computing   212

PAMPA DEBNATH, ARPAN DEYASI,
AND SIDDHARTHA BHATTACHARYYA

Preface

Quantum computer, as the name suggests, principally works on several quantum physical features. These could be used as an immense alternative to today's apposite computers since they possess faster processing capability (even exponentially) than classical computers. The term "quantum computing" is fundamentally a synergistic combination of thoughts from quantum physics, classical information theory, and computer science.

Quantum information processing entails the processing of information represented using qubits, which is the basic element of a quantum computer. Information processing in the quantum domain is centered on qubit encoding of the classical information, using the inherent properties of superposition and coherence followed by some quantum measurement operations to arrive at the classical outcome. In addition, the property of entanglement of the qubits helps in long-haul communication at an enhanced data transfer rate. This increased data transfer rate forms the basis for implementing distributed quantum networks in the near future. In addition, faster quantum communication is imminent due to rapid research on quantum information processing, the possible realization of quantum networks, and quantum internet services. Novel algorithms on quantum key distributions have been proposed to pave the way for a secured and robust communication system, thanks to the progress in research on single photon sources and detectors. Sequential single photon communication leads to quantum cryptography, which not only helps prevent eavesdropping but also forms the backbone for building quantum teleportation networks. Although entangled photon generation at the chip level still remains a challenging proposition, semiconductor-based quantum dot detectors and novel optical fibers can become effective for the practical implementation of encryption algorithms.

In addition to device-level research, scientists have invested their efforts to induce intelligence in quantum information processing in order to make the systems robust, fail-safe, and efficient. Utilization of the basic features of quantum computing in different machine learning algorithms has been explored over the decades, thereby evolving quantum-inspired/quantum computational intelligent algorithms. Starting from evolving quantum neural networks to emulating quantum fuzzy principles to evolving quantum metaheuristics, these have been the trends of research in recent years. The advent of these quantum algorithms has opened up a new era in the field of intelligent information processing, where the principles of quantum mechanics are conjoined successfully to enhance the real-time performance of the existing quantum information processing algorithms.

This volume aims to bring together recent advances and trends in the methodological approaches, theoretical studies, and mathematical and applied techniques related to intelligent quantum information processing and their applications to engineering problems. The scope of the book, in essence, is confined to but not bounded to introducing different novel hybrid quantum computational algorithms for addressing the limitations of the conventional information processing algorithms, including quantum machine learning, quantum key distribution, quantum information processing, quantum encryption algorithms, quantum networks, and quantum knowledge discovery in databases, to name a few. It is also aimed to emphasize the effectiveness of the proposed approaches over the state-of-the-art existing approaches by means of illustrative examples and real-life case studies.

This volume contains ten well-versed chapters, addressing different facets of intelligent quantum information processing.

The exponential rise in the demand for secured information processing requires an alternative approach from the conventional classical world, and the present scientific community believes that one of the most distinctive physical resources of the quantum world is quantum entanglement. It is a unique correlation between the components of quantum multipartite systems, which has never been observed in the classical macroscopic world. It is a tool that not only explains the effectiveness and fruitfulness of quantum information processing compared to its classical counterpart but also becomes significant from the security point of view when quantum communication is considered. Chapter 1 discusses the importance of quantum entanglement as regards to processing of information for secured data transmission. The most trusted and investigated resource to date is quantum entanglement, for the purpose of secured data communication. Similar to energy, quantum entanglement is a physical resource connected to the odd nonclassical connections that can occur between distant quantum systems. Using a pair of entangled quantum systems, it is feasible to carry out computing operations that are impractical for conventional systems. Quantum information theory is a broad study of the information processing capabilities of quantum systems.

At the most microscopic level, all physical systems are governed by the laws of quantum mechanics. Quantum information processing is the study of how information is gathered, transformed, and transmitted at the quantum level—in atoms, ions, photons, elementary particles, and microscopic solid-state systems, which obey fundamental quantum mechanical laws. Chapter 2 highlights the essential tenets of quantum information processing for the next-generation communication system design. Quantum computers, quantum communication channels, and quantum sensors are devices that can attain the ultimate limits of information processing. The laws of quantum mechanics give rise to counterintuitive effects. Quantum information processors use "quantum weirdness" to perform tasks that classical information processors cannot. Quantum computers are conceived to process information stored on atomic, optical, and solid-state systems. They aim to use counterintuitive effects including quantum superposition and entanglement to perform tasks such as quantum simulation, quantum search, and factoring/code-breaking to solve problems that are hard or impossible on conventional classical computers. Quantum communication systems transmit information encoded in individual photons: they exploit the fact that quantum measurement is inevitably stochastic and destructive to enact quantum-encrypted communication whose security is guaranteed by the laws of physics. Quantum sensors and measurement devices operate at the greatest possible sensitivity and precision allowed by physical law: from magnetometers to quantum clocks to advanced gravitational interferometers (e.g., LIGO), quantum metrology supplies the techniques required to push measurement to its ultimate limits.

Extracting and automatically classifying information elements (UIE) such as tables, figures, and equations from (scientific) documents is a challenge in the area of information retrieval. Convolutional neural networks (CNNs) are widely used models in image classification due to their good performance in pattern detection. However, CNNs present some problems in learning large tasks because they depend on having a considerable amount of data for optimal performance. On the contrary, Quantum Convolutional Neural Networks (QCNNs) have been increasingly used because they provide efficient solutions using the concept of quantum computing to improve the performance of a learning model. In Chapter 3, a Hybrid Quantum Convolutional Quantum Neural Network (HQCNN) is analyzed and compared with different state-of-the-art CNNs in the problem of classifying tables extracted from scientific documents. Given the limited access to specialized hardware, in this work, the simulations were carried out using the PennyLine platform and the Multiscale Entanglement Renormalization Ansatz (MERA) model. Results obtained in this work show that HQCNN results in superior performance than a traditional convolutional network compared to the literature.

Quantum computing, a branch of quantum physics, is seen as a promising solution to many of the world's problems. This dynamic research field has gained momentum over the past two decades, driven by the interaction between light and matter. Electromagnetic (EM) waves resonantly interacting with charged particles result in electrodynamics, as described by Maxwell's equations. Maxwell's equations are known for their form-invariance under arbitrary spatial coordinate transformations, enabling effects resulting from a coordinate transformation to be assimilated by the material properties that the EM waves pass through. Transformation optics (TO) applies a coordinate transformation-based approach to create a non-homogeneous and an anisotropic transformation medium, paving the way for unique electromagnetic and optical devices previously thought impossible. TO concepts can also be applied to the regions containing electromagnetic sources, known as source transformations. Chapter 4 delves into the theoretical and mathematical foundations of TO to help readers better understand the concepts behind it. Using this knowledge, this chapter discusses a phased array antenna with new pinwheel-shaped elements that incorporate structural and mechanical constraints using source transformations. Numerical simulations are used to demonstrate the radiation properties of the array, which has the potential to be integrated into conformal arrays for wireless communications, radars, and sensing. This chapter also discusses the future direction of TO-based device design, including the potential use of deep learning (DL), a subfield of machine learning (ML), and artificial intelligence (AI), to predict material properties and to determine design parameters for the desired performance.

The authors present an improved method for quantum machine learning, using a modified Levenberg-Marquardt (LM) method, in Chapter 5. The LM method is a powerful hybrid gradient-based reinforcement learning technique, ideally suited to quantum machine learning, as it only requires knowledge of the final measured output of the quantum computation, not intermediate quantum states, which are generally not accessible without collapsing the quantum state. With this method, the authors have been able to achieve true online training in a quantum system to do a quantum calculation, which has never been done before. The authors demonstrate this using a fundamentally non-classical calculation: estimating the entanglement of an unknown quantum state. Machine learning is applied to learn this algorithm and is demonstrated in simulation and hardware. Results are exhibited for two-, three-, four-, five-, six-, seven-, and eight-qubit systems, in Matlab simulations, and, more importantly, these run on the IBM Qiskit hardware interface as well. With this approach, the quantum system, in a sense, designs its own algorithm. Moreover, the approach enables scaleup, is potentially more efficient, and provides robustness to both noise and decoherence.

When an optimization problem needs to be solved, it is possible to use some classical methodology, such as the gradient-guided or exhaustive search technique. However, to be able to use any of them, it is necessary to have an extra knowledge of the problem. Evolutionary algorithms have been tested and applied to a wide variety of problems of this type, being a very good search strategy having the characteristic of not requiring extra information from the problems like those previously mentioned. One type of evolutionary computation is Genetic Algorithms (GAs), which are based on the mechanics of natural selection proposed by Charles Darwin. In a GA, a population of possible solutions is generated, and later, this is evolved through genetic operators (crossover, mutation, and selection) to generate a new one with better individuals (new possible solutions) that are close to the sought solution. Since 1996, a modification to GAs was presented by Narayanan and Moore, employing quantum mechanisms where the Schrödinger equation is essential to bring an initial state to a final state, the Quantum Genetic Algorithms (QGAs). Here, the concept of qubit arises, which aims to store the minimum unit of information of the quantum states, being $|0\rangle$ and $|1\rangle$. The advantage of this is that now it is also possible to have a state superposition of the two previous states $|\psi\rangle = \alpha|0\rangle + \beta|1\rangle$, increasing the space of solutions. Similar to GAs, in QGAs, there are quantum crossover and quantum mutation (Q-gates), genetic operators that help to evolve the initial population toward the optimal value sought. In Chapter 6, a QGA is used to fit a curve on an image, the Major Temporal Arcade (MTA) vein in fundus images, which has been segmented previously. The aim is to determine the coefficients and the degree of the polynomials that, in linear superposition, fit the curve as best as possible, finally having a functional expression that can be manipulated in order to extract as much information as possible and help in the diagnosis and treatment of diseases related to the retina.

Quantum computing, which is the most famed topic in this era, a combination of quantum mechanics from the 1900s and computing point of view, takes a huge place in the quantum realm. Starting from the foundation of abacus from 2700 BC to 2300 BC, we humans always try to compute in different ways; the most modern ways of doing it come with the binary system. Bit is the smallest unit of a binary system and can be in two states: either 0 or 1. On the contrary, quantum computing deals with the "quantum bit" or "qubit"; possible outcomes from the qubits are neither 0 nor 1. It will always be a combination of 0 and 1, which is based on the superposition principle from quantum physics. In Chapter 7, the details of quantum logic gates are discussed from a circuit design perspective, which will be achievable for future realization of quantum computers. The architecture and the corresponding formation of basic quantum circuits are discussed, along with methods of quantum computing. How the technique differs from conventional approaches is also highlighted. The work may be directed toward future hardware realization of quantum information processing.

Quantum computing is a rapidly evolving industrial technology that employs quantum mechanics laws to address issues that traditional archaic computers find impossible to solve. It is a computation method that uses phenomena such as superposition, interference, and entanglement. It is one of the most effective theories of the twentieth century, contributing to the progressive growth of scientific investigations. In this modern-day world, there have been certain advancements in quantum computers that have encouraged researchers to opt for a number of algorithms. There are certain drawbacks in the architecture, mainly the presence of certain physical constraints that restrict the logical qubits from getting mapped and converted to physical qubits. Chapter 8 puts forward the modern-day trends in the field of quantum computing and focuses on the various restrictions faced in building a device following the principles of quantum computing.

Chapter 9 deals with fundamental concepts and corresponding algorithms related to quantum microwave propagation and its application in various sensing and communication networks. Propagation in both unbounded and bounded media is briefly mentioned with a detailed elaboration on qubits. The role of qubits as resonators is analyzed along with their physical representation. A few adopted algorithms are mentioned, which are nowadays essentially used for realizing quantum logic gates for sequential circuits. Both spin qubits and superconducting qubits are discussed at the end of this chapter to emphasize their future potential application areas.

Finally, Chapter 10 concludes the findings of the volume and future research directions.

This volume is meant for undergraduates and postgraduates of information science, computer science, electronics, and communication engineering for some part of their curriculum. The volume would also benefit aspiring and senior researchers to carry out further exploration of the application areas pertaining to this upcoming technology. The volume is also intended to benefit the faculty of relevant disciplines in premier institutes.

About the Editors

Siddhartha Bhattacharyya did his Bachelor in Physics, Bachelor in Optics and Optoelectronics, and Master in Optics and Optoelectronics from the University of Calcutta, India, in 1995, 1998, and 2000, respectively. He completed his PhD in Computer Science and Engineering from Jadavpur University, India, in 2008. He completed a habilitation thesis from VSB Technical University of Ostrava, Ostrava, Czech Republic, in 2023. He is the recipient of the University Gold Medal from the University of Calcutta for his Master. He is the recipient of several coveted awards, including the Distinguished HoD Award and Distinguished Professor Award conferred by the Computer Society of India, Mumbai Chapter, India, in 2017; the Honorary Doctorate Award (D. Litt.) from the University of South America; and the South East Asian Regional Computing Confederation (SEARCC) International Digital Award ICT Educator of the Year in 2017. He has been appointed as the ACM Distinguished Speaker for the tenure of 2018–2020. He was inducted into the People of ACM Hall of Fame by ACM, the USA, in 2020. He has been appointed as the IEEE Computer Society Distinguished Visitor for the tenure of 2021–2023. He has been elected as a full foreign member of the Russian Academy of Natural Sciences (RANS) and the Russian Academy of Engineering (REA). He has been elected a full fellow of the Royal Society for Arts, Manufacturers and Commerce (RSA), London, UK. He is currently serving as a Senior Researcher in the Faculty of Electrical Engineering and Computer Science of VSB Technical University of Ostrava, Czech Republic. He is also serving as the Scientific Advisor of Algebra University College, Zagreb, Croatia. Prior to this, he served as the Principal of Rajnagar Mahavidyalaya, Rajnagar, Birbhum. He served as a Professor in the Department of Computer Science and Engineering of Christ University, Bangalore. He was the Principal of RCC Institute of Information Technology, Kolkata, India, from 2017 to 2019. He has also served as a Senior Research Scientist in the Faculty of Electrical Engineering and Computer Science of VSB Technical University of Ostrava, Czech Republic (2018–2019). Prior to this, he was a Professor of Information Technology at RCC Institute of Information Technology, Kolkata, India. He served as the Head of the

Department from March 2014 to December 2016. Prior to this, he was an Associate Professor of Information Technology at RCC Institute of Information Technology, Kolkata, India, from 2011 to 2014. Before that, he served as an Assistant Professor in Computer Science and Information Technology at the University Institute of Technology, the University of Burdwan, India, from 2005 to 2011. He was a Lecturer in Information Technology at Kalyani Government Engineering College, India, during 2001–2005. He is a co-author of 6 books and the co-editor of 100 books and has more than 400 research publications in international journals and conference proceedings to his credit. He has 2 patent cooperation treaties and 19 patents to his credit. He has been a member of the organizing and technical program committees of several national and international conferences. He is the Founding Chair of ICCICN 2014, ICRCICN (2015, 2016, 2017, 2018), and ISSIP (2017, 2018) (Kolkata, India). He was the General Chair of several international conferences such as WCNSSP 2016 (Chiang Mai, Thailand), ICACCP (2017, 2019) (Sikkim, India), ICICC 2018 (New Delhi, India), and ICICC 2019 (Ostrava, Czech Republic). He is the Associate Editor of several reputed journals, including *Applied Soft Computing, IEEE Access, Evolutionary Intelligence*, and *IET Quantum Communications*. He is the editor of the *International Journal of Pattern Recognition Research* and the Founding Editor-in-Chief of the *International Journal of Hybrid Intelligence*, Inderscience. He has guest-edited several issues with several international journals. He is serving as the Series Editor of IGI Global Book Series Advances in Information Quality and Management (AIQM), De Gruyter Book Series(s) Frontiers in Computational Intelligence (FCI) and Intelligent Biomedical Data Analysis (IBDA), CRC Press Book Series(s) Computational Intelligence and Applications, Quantum Machine Intelligence, Intelligent Data Driven Technology for Sustainable Environment, and Advances in Disruptive Technologies and Generative Artificial Intelligence, Elsevier Book Series Hybrid Computational Intelligence for Pattern Analysis and Understanding and Springer Tracts on Human Centered Computing. His research interests include hybrid intelligence, pattern recognition, multimedia data processing, social networks, and quantum computing. He is a life fellow of the Optical Society of India (OSI), India; a life fellow of the International Society of Research and Development (ISRD), UK; a fellow of the Institution of Engineering and Technology (IET), UK; a fellow of the Institute of Electronics and Telecommunication Engineers (IETE), India; and a fellow of the Institution of Engineers (IEI), India. He is also a senior member of the Institute of Electrical and Electronics Engineers (IEEE), USA; the International Institute of Engineering and Technology (IETI), Hong Kong; and the Association for Computing Machinery (ACM), USA. He is the Founding President of the Asia-Pacific Artificial Intelligence Association (AAIA), Kolkata, and the Chairman of the IEEE Computational Intelligence Society, Kolkata Chapter. He is a life member of the Cryptology Research Society of India (CRSI), Computer Society of India (CSI), Indian Society for Technical Education (ISTE), Indian Unit for Pattern Recognition and Artificial Intelligence (IUPRAI), Center for Education Growth and Research (CEGR), Integrated Chambers of Commerce and Industry (ICCI), and Association of Leaders and Industries (ALI). He is a member of the Institution of Engineering and Technology (IET), UK; International Rough Set Society, International Association for Engineers (IAENG), Hong Kong; Computer Science Teachers Association (CSTA), USA; International Association of Academicians, Scholars, Scientists and Engineers (IAASSE), USA; Institute of Doctors Engineers and Scientists (IDES), India; the International Society of Service Innovation Professionals (ISSIP); and the Society of Digital Information and Wireless Communications (SDIWC). He is also a certified Chartered Engineer of the Institution of Engineers (IEI), India. He is on the Board of Directors of the International Institute of Engineering and Technology (IETI), Hong Kong.

Iván Cruz-Aceves received an MS degree in Computer Science in 2009 from the Leon Institute of Technology and a PhD in Electrical Engineering from the University of Guanajuato with Summa Cum Laude distinction in 2014. He joined the Department of Computer Science at the Centre for Research in Mathematics (CIMAT) in 2014 as part of the program "Investigadores por México" at the National Council for Science and Technology (CONACYT). He is currently a full-time researcher of Computer Science, and he is the author of more than 30 international conference papers, 35 papers in journals (JCR), international book chapters, and 10 transfer technologies to the medical area, involving medical image processing and analysis. His areas of interest are Artificial Intelligence focused on stochasticity and Evolutionary Computation with applications to medical image and video processing and analysis.

Pampa Debnath is presently working as an Assistant Professor in the Department of Electronics and Communication Engineering at RCC Institute of Information Technology, Kolkata, India. She has more than 16 years of professional teaching experience in academics. Her research interest covers the area of microwave devices, high-frequency antennas, SIW, and RGW-based circuits. She has already served as the technical chair and session chair of several IEEE, Springer, and other international conferences and coordinated several faculty development programs, workshops, laboratory and industrial visits, seminars, and technical events under the banner of IEEE and the Institution of Engineers, Kolkata section, India. She is a reviewer of a few journals of repute and some IEEE, Springer, and other national and international conferences. She is a member of the Institute of Engineers (IE), the Institution of Electronics and Telecommunication Engineers (IETE), the Indian Society for Technical Education (ISTE), and the International Association for Engineers (IAENG).

Arpan Deyasi is presently working as an Associate Professor in the Department of Electronics and Communication Engineering in RCC Institute of Information Technology, Kolkata, India. He has more than 17 years of professional experience in academia and industry. His work spans around in the field of semiconductor nanostructure and semiconductor photonics. He has published more than 200 peer-reviewed research papers and edited 8 books. He is associated with different international and national conferences in various aspects and is also Guest Editor of a few renowned journals. He is a senior member of IEEE, Vice-Chair of IEEE Electron Device Society (Kolkata Chapter), and a member of IE(I), Optical Society of India, IETE, ISTE, ISVE, and so on.

Rajarshi Mahapatra received a PhD degree in electronics and electrical communication engineering from the Indian Institute of Technology Kharagpur, Kharagpur, India, and a postdoctoral degree from the CEA-LETI, Grenoble, France. In his postdoctoral research, he was engaged in FP7 Call4 BeFEMTO and Greentouch projects. He is presently serving as an Associate Professor with the Department of Electronics and Communication Engineering, Dr. SPM IIIT, Naya Raipur, Chhattisgarh, India. He also served as the Dean (Academics) of IIITNR. Earlier, he had worked for Collins Aerospace, Hyderabad, on software-defined radio and electronic warfare.

He has worked extensively in the domain of physical layer design and analysis of a wireless communication system. He worked in the fields of cognitive radio, 5G and 6G communication, heterogeneous wireless communication, molecular communication, and energy-efficient communication. His team designed and developed software-defined radio and direction-finding systems for EW applications in Collins Aerospace industry. He has about 18 years of teaching, research, and industry experience. He has guided PhD scholars in the area of wireless communication. He has published several research papers in various refereed journals and IEEE journals. He is a regular reviewer of premier IEEE Transactions and other peer-reviewed journals and IEEE conferences. He has been organizing many workshops on 5G in recent years. He has also developed several high-value research labs, including the 5G test bed. He has successfully completed and undertaken high-value sponsored projects in the field of communication systems.

He has been awarded a national scholarship, an MHRD scholarship for research, and an EU-FP7 fellowship for a European project. He is a senior member of IEEE and a member of the Communication Society. His research interests include 5G and beyond communication, machine learning for communication, molecular communication, intelligent reflecting surfaces, and optical access networks.

Contributors

Maria-Susana Avila-Garcia
Departamento de Estudios
  Multidisciplinarios
División de Ingenierías
Universidad de Guanajuato
Av. Universidad S/N, Col. Yacatitas
Yuriria, GTO, Mexico

Elizabeth C. Behrman
Department of Mathematics,
  Statistics and Physics
Wichita State University
Wichita, Kansas

Joy Bhattacharjee
Department of System Engineering
Pixel Consultancy
Kolkata, India

Siddhartha Bhattacharyya
Department of Electrical Engineering
  and Computer Science
VSB Technical University of Ostrava
Czech Republic and Algebra
  University College
Zagreb, Croatia

Iván Cruz-Aceves
CONACYT
Centro de Investigación en
  Matemáticas (CIMAT)

A.C., Jalisco S/N, Col. Valenciana
Guanajuato, Mexico

Pampa Debnath
Department of Electronics
  and Communication
  Engineering
RCC Institute of Information
  Technology (RCCIIT)
Kolkata, India

Shuvashis Dey
Department of Electrical and
  Computer Engineering
North Dakota State University
Fargo, North Dakota

Arpan Deyasi
Department of Electronics and
  Communication Engineering
RCC Institute of Information
  Technology (RCCIIT)
Kolkata, India

Erick Franco-Gaona
Departamento de Estudios
  Multidisciplinarios
División de Ingenierías
Universidad de Guanajuato
Av. Universidad S/N, Col. Yacatitas
Yuriria, GTO, Mexico

Ratneswar Ghosh
Department of Electrical Engineering
Techno International New Town
Kolkata, India

Arturo Hernández-Aguirre
Departamento de Computación
Centro de Investigación en
    Matemáticas (CIMAT)
A.C., Jalisco S/N, Col. Valenciana
Guanajuato, Mexico

Martha Alicia Hernández-González
Unidad Médica de Alta
    Especialidad (UMAE)
Hospital de Especialidades No. 1.
    Centro Médico Nacional del Bajio
IMSS, Blvd. Adolfo López Mateos
    esquina Paseo de los Insurgentes
    S/N, Col. Los Paraisos
León, Mexico

Eric Jahns
Department of Computer Science
    and Computer Engineering
University of Wisconsin–La Crosse
La Crosse, Wisconsin

Luis Miguel López-Montero
Unidad Médica de Alta
    Especialidad (UMAE)
Hospital de Especialidades No. 1.
    Centro Médico Nacional del Bajio
IMSS, Blvd. Adolfo López Mateos
    esquina Paseo de los Insurgentes
    S/N, Col. Los Paraisos
León, Mexico

Dipankar Mitra
Department of Computer Science
    and Computer Engineering
University of Wisconsin-La Crosse
La Crosse, Wisconsin

Krishnanjan Mukherjee
Department of Electrical
    Engineering
Techno International New Town
Kolkata, India

Sayan Roy
Department of Electrical and
    Computer Engineering
Division of Energy, Matter, and
    Systems
University of Missouri–Kansas City
Kansas City, Missouri

Soumen Santra
Department of Computer
    Applications
Techno International New Town
Kolkata, India

Sweta Sharma
System Engineering
Accenture Pvt Ltd
Kolkata, India

José Alfredo Soto-Álvarez
Departamento de Computación
    Centro de Investigación en
    Matemáticas (CIMAT)
A.C., Jalisco S/N, Col. Valenciana
Guanajuato, Mexico

James E. Steck
Department of Aerospace
    Engineering
Wichita State University
Wichita, Kansas

Nathan L. Thompson
Department of Mathematics,
    Statistics and Physics,
Wichita State University
Wichita, Kansas

Chapter 1

The Role of Quantum Entanglement in Information Processing for Secured Data Transmission

Arpan Deyasi, Pampa Debnath, and Siddhartha Bhattacharyya

1.1 INTRODUCTION

Using the rules of quantum physics to tackle issues that are hard for conventional, antiquated computers to solve, quantum computing is a rapidly developing industrial technology. It is a technique for computing that makes use of entanglement, interference, and superposition [1]. In order to process data accumulated on optical, atomic, and solid-state systems, quantum computers are developed. They are designed to use counterintuitive phenomena such as quantum superposition and entanglement to carry out operations [2] like quantum simulation, quantum search, and factoring/code-breaking, which are difficult or impossible to carry out on conventional classical computers. Coherent quantum connections between light and matter, as well as quantum error correction and communication, are needed, for the creation of large-scale quantum computers and a quantum internet. A turning point has been accomplished in the advancement of ever-larger-scale quantum computing [3]. Despite an exponential progress in quantum technology during the past two decades, in general-purpose quantum computers, the density of quantum bits did not quadruple every two years. In physical structure, the amount of information that can be collected, as well as the rate at which those bits can flip, is governed by the laws of quantum physics.

It is not an issue how far away two arrangements are, when they are in the quantum entangled state; studying something about one instantaneously discloses something about the other. The fact that this phenomenon challenges the assumption that no information can be sent quicker than the speed of light confused Einstein, who described it as "a spooky action at a distance." However, further research must be done to demonstrate entanglement by means of photons and electrons. If an entangled qubit's state changes in a quantum computer, the related qubit's state will likewise change instantly [4]. Thus, entanglement develops the processing velocity of quantum computers. Even if one qubit can provide details about many other processes [5], twice the number of entangled bits or qubits does not always result in a quadrupling of the processes.

DOI: 10.1201/9781003373117-1

1.2 QUANTUM INTERFERENCE

The experiment with two slits through which a stream of particles passes and creates an interference pattern that resembles waves on a screen on the other side is a famous illustration of quantum superposition. The fundamental characteristic of quantum interference is that even when there is only one particle present in the device at any given moment, an interference pattern can still arise [6]. It is necessary for quantum interference that the experiment should be conducted in a manner that it is impossible to determine, even in theory, where journey of particle through both the slits are considered.

In order to explain quantum interference, the particle is in a superposition of the two experimental trails: channel through the upper slit and channel through the lower slit. A quantum bit can also be superposed as |0⟩ and |1⟩. The idea is the same; however, experiments in quantum information processing often employ interferometers rather than double slits. Amounts of electrons, neutrons, photons, and atoms have all been used to study single-particle quantum interference thus far. Figure 1.1 shows quantum interference between two photons emitted from two quantum dots, respectively.

Quantum interference comes in two flavors: constructive interference and destructive interference. When two in-phase waves that peak at the same moment interact positively, the resultant wave has a peak that is twice as high. On the contrary, two waves that are out of phase peak at opposing times and interact destructively; the resulting wave is entirely flat. With a larger peak for constructive interference or a lower peak for destructive interference, all other phase differences will provide outcomes that fall somewhere in the middle [7].

The wave-particle duality is the fundamental cause of quantum interference. Particles at the subatomic level exhibit wavelike characteristics [8]. These wavelike characteristics are frequently related to position; for instance, the area an electron could be in relation to a nucleus. Because of this, electron orbitals are represented as probability clouds rather than as planet-like orbits around the sun. However, this locational uncertainty also applies to energy levels and the potential electron orbital.

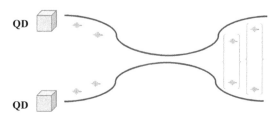

Figure 1.1 Quantum interference between two photons.

1.3 QUANTUM SUPERPOSITION

A physical system that exists in several states simultaneously depends on a certain set of solutions, and this property of quantum mechanics is known as superposition. All feasible solutions, usually referred to as Hilbert space, are the set of solutions that are most frequently employed. The mathematical representation of all the various states a system might have in quantum physics is called the Hilbert space [9]. Given that there are two potential spin directions for my system, which is a spinning electron, its Hilbert space is spin up and spin down. Since it was first offered eighty years ago, the "Schrodinger's cat" thought experiment has generated a great deal of curiosity. It has been possible to realize superposition states for photons, electrons, atoms, and certain compounds [10]. In photosynthetic systems, wavelike energy transmission through quantum coherence has been reported [11, 12]. Figure 1.2 represents quantum superposition between two classical bits, which results in qubits.

1.4 QUANTUM ENTANGLEMENT

Quantum interference leads to quantum entanglement. It explains how two subatomic particles may be closely connected to each other while being billions of light-years apart. When two qubits are entangled, they are connected in a unique way. The results of measurements will reveal the entanglement. The results of the measurements performed on each qubit might be a 0 or a 1. However, the results of one qubit's measurement will always be correlated with those of the other qubit's measurement [13]. Even if the particles are far apart from one another, this is always the case. In order to generate entangled systems, one must first create a pair of electrons with opposing spins, as required by the Pauli exclusion principle, while maintaining the quantum uncertainty

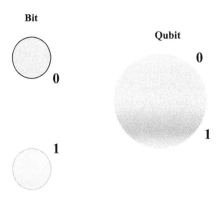

Figure 1.2 Quantum superposition between two classical bits.

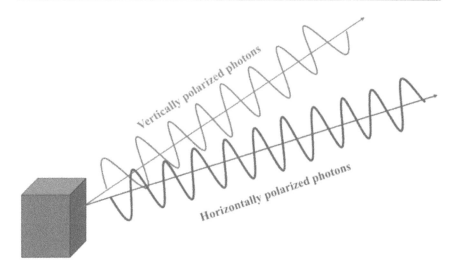

Figure 1.3 Quantum entanglement between vertically and horizontally polarized photon waves.

of each particle's real spin. When a pair of particles is separated, even by a great distance, and one particle's spin is measured, the second particle's spin will naturally resolve itself in the opposite direction [14]. A pair of entangled particles is employed in quantum teleportation to instantly transport data on a third item from one location to another [15]. Figure 1.3 represents entanglement between two photon waves mutually orthogonal to each other.

1.5 QUANTUM COMMUNICATION

All three of these quantum communication applications have made substantial strides during the past ten years. In the past two years, the theoretical limit of these channels has finally been determined; though people still don't know how to really get there, they are aware of the end state. Linked quantum computers might permit consumers to communicate combined instruction in a reliable manner, varying from quantum secret allotment and quantum resolution creating to fraud-free quantum voting, in accordance with the notion of distributed quantum computation across a quantum internet [16]. Like how the conformist, classical internet has permitted a broad range of applications that go beyond the usage of data, the quantum internet has the potential to transform how citizens and associations interact and struggle, improving faith while defending solitude.

Quantum communication has several uses and significant ramifications. Quantum data locking, quantum secret sharing, and quantum key distribution are just a few examples of the physically secure communication protocols

Figure 1.4 Quantum communication between two stations.

that may be created thanks to quantum physics. Innovative approaches to computer security include quantum data locking and "blind" quantum computation, which rely on the ideology of quantum mechanics to protect stored data and make a quantum computer inaccessible to outside users. They significantly reduce the attack resistance of quantum computers connected to a quantum internet [17]. Figure 1.4 describes the inherent zero loss property (w.r.t information transmission) when quantum communication replaces the classical mode.

A quantum internet would enable distributed quantum information processing, precise sensing and navigation, and long-distance secure communication, but these are not possible in a conventional setting. Numerous varieties of quantum memory and information processors coupled with optical lines will certainly make up a quantum internet [18]. It will include a variety of unique quantum memory and sensor technologies, including cold atoms, solid state, photonic, and superconducting, as was previously stated [19]. This heterogeneity resembles a crucial element of the modern "classical internet of things," as it links several information processing systems, each of which has advantages and disadvantages of its own.

1.6 QUANTUM INFORMATION PROCESSING

The major areas of quantum information processing include information processing and computation based on quantum physics. Quantum computers are not constrained to two states, unlike conventional digital computers that encode data in binary digits (bits). Quantum bits, or qubits, which may exist in superposition, are used to encode information. Atoms, ions, photons, or electrons, along with the appropriate control mechanisms, can be used to create qubits, which can function as a computer's processor and memory [20]. These numerous states may all exist concurrently in a quantum computer, which gives them an intrinsic parallelism.

Researchers have made slow but considerable progress in creating quantum algorithms and hardware since the quantum-mechanical computer was first imagined by Richard Feynman. Turing machines are thought to be incapable of doing operations like factorization in polynomial time, while quantum machines can. Deutsch and David proposed that "Deutsch-Jozsa algorithm" [21] was introduced by Richard Jozsa as an example of how a

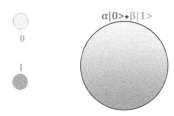

Figure 1.5 Quantum nature added with physical information.

quantum algorithm may finish a job more quickly than a traditional one. The most popular quantum algorithm, Shor's algorithm [22], may factor a prime number with an integer 10 times faster than the most excellent conformist approach. The experimental hardware implementation of the Controlled NOT (CNOT) gate was proposed by Cirac and Zoller [23]. There have been advancements in quantum computing, particularly in recent years. Quantum computing has made strides in all areas of the stack, considering algorithm, structural design, and hardware, particularly in the recent several years. The most noteworthy of these is Google's proof of quantum supremacy [24]. Figure 1.5 shows the effect of quantum nature when mixed with physical information.

The worldwide quantum computer, a quantum mechanical equivalent of the pervasive traditional computer, is the ultimate tool for processing quantum information. Manin and Feynman [25, 26], who noted that an essentially quantum mechanical computer would be well-suited to the modeling of quantum systems, a challenge that appears to be extremely tough for classical computers, first proposed the fundamental concept of a quantum computer in the early 1980s. In 1985, Deutsch [27] recommended the thought of a quantum computer as a general-purpose computing tool and offered the initial illustration of a problem that a quantum computer might tackle more quickly than a conventional computer. After Deutsch's work, there was a continuous progression of improvements that culminated in 1994 with Shor's discovery of effective quantum algorithms for computing discrete logarithms and factoring numbers [28]. Shor's technique shows that building a quantum computer might have substantial practical implications since the security of most modern cryptosystems is predicated on the alleged difficulty of factoring. The essential component based on which the framework of information processing stands is qubit. This is discussed in the next section.

1.7 QUBIT

The fundamental data storage unit in a quantum computer (QC) is the quantum bit or qubit, taking the place of the bit in a traditional computer. Both the ground and the excited states of the qubit are possible at once. Each qubit's two logical states must be translated onto the eigenstates of an appropriate physical system. A spin degree of freedom, which can retain quantum information, a little amount for exceptionally extended period [29], is the foundation of a spin qubit. Be aware that there are several other qubit instances, such as the two distinct photon polarizations, the two energy eigenstates of an electron, and so on. Due to two unique qubit characteristics, a quantum computer is primarily dissimilar from a traditional computer [30]. Quantum superposition, or the linear combination of potential configuration, is the first characteristic. Quantum entanglement is the second.

The following five conventional conditions, known as the DiVincenzo criteria after theoretical physicist David P. DiVincenzo [31], should be met by the qubits in a quantum computer:

1. A physically scalable system
2. The ability to set the qubits' state
3. Worldwide place of quantum gates
4. Coherence period larger than the gate operation time
5. Capability for measurement particular to qubits [32]

1.8 SECURED DATA COMMUNICATION

A photon of light, which is generally employed as an information carrier in fiber optic cables, can be any combination of 1s and 0s in quantum communication. According to conventional communication, the photon could only be either a '1' or a '0', not both. A quantum bit of information, or qubit, in quantum communication is represented by a photon's capacity to combine 1s and 0s. The fact that qubits are "super fragile" is a key benefit for cybersecurity while delivering information. Accordingly, if a hacker tries to steal the data, it is noticed and collapses to a '1' or a '0'. The qubit's collapse to a simpler '1' or '0' gives a clear indication that the message was being attempted to be insinuated [33]. The information conveyed is essentially corrupted by the collapse of the quantum state that is essential to the qubit, preventing the "would-be" hacker from exploiting it for any nefarious purposes.

1.9 QUANTUM KEY DISTRIBUTION

A mix of more conventional encrypted communication with quantum communication is known as quantum key distribution (QKD). Informational bits are encrypted and transferred over the network in QKD. The encryption key is encoded into qubits and is transferred through a particular QKD channel, which is where quantum physics is used [34]. To standardize different QKD techniques, multiple protocols have been established, while the technology is still in its relative infancy. An example that simplifies the protocol is to transfer secure data to another machine. The data sender will provide an encryption key that will be used to decrypt the message itself [35]. The key is protected with qubits.

The recipient is then supplied the data and the encryption key, and using a process known as key sifting, it can be established that the key is the same for both the parties. Using the verified encryption key, the data may then be decoded. Quantum-resistant cryptography is another name for post-quantum cryptography (PQC) [36, 37], which consists of fresh methods for mathematical encoding that are impervious to Shor's algorithm's assault as well as any potential future quantum algorithms.

1.10 QUANTUM INTERNET

By coupling operational quantum computers with quantum communication channels, a quantum communication network enables dispersed quantum information processing. It provides information processing skills that are not possible with conventional computing techniques. Its use examples include developments in quantum sensor networks, distributed quantum computing, and long-haul secure communication [38].

The current internet and the quantum internet are intended to complement each other. In conjunction with the quantum bits (officially known as qubits), consumer-oriented social media images, music videos, and a lot of non-sensitive business information will still be sent as classical bits [39]. In addition to metrology and quantum computation, QKD—supposed to be its

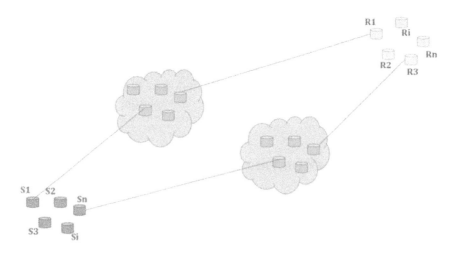

Figure 1.6 Quantum network formed using quantum internet.

best-known application—could draw businesses eager to keep sensitive data safe and information moving. Quantum network, formed using quantum internet, is exhibited in Figure 1.6.

The first multinode quantum networks are anticipated to emerge over the next several years, even if it is difficult to foresee what the precise physical elements of a future quantum internet will be [38]. This innovation offers the thrilling chance to put to the test all the concepts and features that, up until now, have only been imagined on paper, and it might very well be the beginning of a future, extensive quantum internet.

REFERENCES

[1] Z. Meng, "Review of quantum computing", 13th IEEE International Conference on Intelligent Computation Technology and Automation, 2020, Xi'an, China

[2] R. Rietsche, C. Dremel, S. Bosch, L. Steinacker, M. Meckel, J.-M. Leimeister, "Quantum computing", Electronic Markets, Vol. 22, pp. 2525–2536, 2022

[3] A. Erhard, J. J. Wallman, L. Postler, M. Meth, R. Stricker, E. A. Martinez, P. Schindler, T. Monz, J. Emerson, R. Blatt, "Characterizing large-scale quantum computers via cycle benchmarking", Nature Communications, Vol. 10, No. 5347, 2019

[4] A. K. Pati, S. Braunstein, "Role of entanglement in quantum computation", Journal of the Indian Institute of Science, Vol. 89, No. 3, pp. 295–302, 2009

[5] M. Zidan, "A novel quantum computing model based on entanglement degree", Modern Physics Letters B, Vol. 34, No. 35, p. 2050401, 2020

[6] J. Lim, S. Kumar, Y. S. Ang, L. K. Ang, L. J. Wong, "Quantum interference between fundamentally different processes is enabled by shaped input wavefunctions", Advanced Science, Vol. 10, No. 10, p. 2205750, 2023

[7] K. Qian, K. Wang, L. Chen, Z. Hou, M. Krenn, S. Zhu, X.-S Ma, "Multiphoton non-local quantum interference controlled by an undetected photon", Nature Communications, Vol. 14, No. 1480, 2023

[8] M. Melucci, "An investigation of quantum interference in information retrieval", IRFC 2010: Advances in Multidisciplinary Retrieval, Vol. 6107, pp. 136–151, 2010

[9] D. J. Wineland, "Nobel lecture: Superposition, entanglement, and raising Schrödinger's cat", Review of Modern Physics, Vol. 85, No. 3, pp. 1103–1114, 2013

[10] K. Hornberger, S. Gerlich, P. Haslinger, "Quantum interference of clusters and molecules", Review of Modern Physics, Vol. 84, No. 1, p. 157, 2012

[11] T. Brixner, J. Stenger, H. M. Vaswani, M. Cho, R. E. Blankenship, G. R. Fleming, "Two-dimensional spectroscopy of electronic couplings in photosynthesis", Nature, Vol. 434, pp. 625–628, 2005

[12] G. S. Engel, T. R. Calhoun, E. L. Read, T.-K. Ahn, T. Mančal, Y.-C. Cheng, R. E. Blankenship, G. R. Fleming, "Evidence for wavelike energy transfer through quantum coherence in photosynthetic systems", Nature, Vol. 446, pp. 782–786, 2007

[13] B. Wong, "On quantum entanglement", International Journal of Automatic Control System, Vol. 5, No. 2, pp. 1–7, 2019

[14] N. Zou, "Quantum entanglement and its application in quantum communication", Journal of Physics: Conference Series, Vol. 1827, p. 012120, 2021

[15] L. Hadjiivanov, I. Todorov, "Quantum entanglement", Bulgarian Journal of Physics, Vol. 42, pp. 128–142, 2015

[16] N. Gisin, R. T. Thew, "Quantum communication technology", Electronics Letters, Vol. 46, No. 14, pp. 965–967, 2010

[17] X. Waintal, "What determines the ultimate precision of a quantum computer", Physical Review A, Vol. 99, p. 042318, 2019

[18] N. Gisin, R. T. Thew, "Quantum communication technology", Electronics Letters, Vol. 46, No. 14, pp. 965–967, 2010

[19] J. Chen, "Review on quantum communication and quantum computation", Journal of Physics: Conference Series, Vol. 1865, No. 022008, 2021

[20] I. B. Djordjevic, "Quantum information processing fundamentals", Quantum Information Processing, Quantum Computing, and Quantum Error Correction, chapter 3, Academic Press, pp. 125–158, 2021

[21] D. Deutsch, R. Jozsa, "Rapid solution of problems by quantum computation", Proceedings of the Royal Society of London. Series A: Mathematical and Physical Sciences, Vol. 439, No. 1907, pp. 553–558, 1992

[22] P. W. Shor, "Polynomial-time algorithms for prime factorization and discrete logarithms on a quantum computer", SIAM Review, Vol. 41, No. 2, pp. 303–332, 1999

[23] J. I. Cirac, P. Zoller, "Quantum computations with cold trapped ions", Physical Review Letters, Vol. 74, No. 20, p. 4091, 1995

[24] F. Arute, K. Arya, R. Babbush, D. Bacon, J. C. Bardin, R. Barends, R. Biswas, S. Boixo, F. G. S. L. Brandao, D. A. Buell, "Quantum supremacy using a programmable superconducting processor", Nature, Vol. 574, No. 7779, pp. 505–510, 2019

[25] F. E. Magniez, M. Santha, M. Szegedy, "Quantum algorithms for the triangle problem", Siam J. Computer: C, Vol. 37, No. 2, pp. 413–424, 2007

[26] R. P. Feynman, "Simulating physics with computers", International Journal of Theoretical Physics, Vol. 21, pp. 467–488, 1982

[27] D. Deutsch, "Quantum theory, the Church-Turing principle, and the universal quantum computer", Proceeding of Royal Society of London A, Vol. 400, pp. 97–117, 1985

[28] P. W. Shor, "Algorithms for quantum computation: Discrete logarithms and factoring", Proceedings of 35th IEEE Symposium on Foundations of Computer Science, IEEE Press, pp. 124–134, 1994

[29] J. J. Pla, K. Y. Tan, J. P. Dehollain, W. H. Lim, J. J. Morton, D. N. Jamieson, A. S. Dzurak, and A. Morello, "A single-atom electron spin qubit in silicon", Nature, Vol. 489, No. 7417, p. 541, 2012

[30] N. T. Bronn, B. Abdo, K. Inoue, S. Lekuch, A. D. Corcoles, J. B. Hertzberg, M. Takita, L. S. Bishop, J. M. Gambetta, and J. M. Chow, "Fast, high-fidelity readout of multiple qubits", Journal of Physics: Conference Series, Vol. 834, p. 012003, 2017

[31] D. P. DiVincenzo, "The physical implementation of quantum computation", Fortschritte der Physik, Vol. 48, No. 911, pp. 771–783, 2000

[32] S. Bravyi, O. Dial, J. M. Gambetta, D. Gil, Z. Nazario, "The future of quantum computing with superconducting qubits", Journal of Applied Physics, Vol. 132, p. 160902, 2022

[33] V. Thayananthan, A. Albeshri, "Big data security issues based on quantum cryptography and privacy with authentication for mobile data center", Procedia Computer Science, Vol. 50, pp. 149–156, 2015

[34] O. K. Jasim, S. Abbas, E.-S. M. El-Horbaty, A.-B. M. Salem, "Quantum key distribution: Simulation and characterizations", Procedia Computer Science, Vol. 65, pp. 701–710, 2015

[35] X. Zheng, Z. Zhao, "Quantum key distribution with two-way authentication", Optical and Quantum Electronics, Vol. 53, No. 304, 2021

[36] A. I. Nurhadi, N. R. Syambas, "Quantum Key Distribution (QKD) protocols: A survey", 4th IEEE International Conference on Wireless and Telematics, 2018, Nusa Dua, Bali, Indonesia

[37] M. Kumar, P. Pattnaik, "Post Quantum Cryptography (PQC)—An overview", IEEE High Performance Extreme Computing Conference, 2020, Waltham, MA, USA

[38] S. Wehner, D. Elkouss, R. Hanson, "Quantum internet: A vision for the road ahead", Science, Vol. 362, No. 6412, 2018

[39] A. Singh, K. Dev, H. Siljak, H. D. Joshi, M. Magarini, "Quantum internet—applications, functionalities, enabling technologies, challenges, and research directions", IEEE Communications Surveys & Tutorials, Vol. 23, No. 4, pp. 2218–2247, 2021

Chapter 2

Quantum Information Processing for Next-Generation Communication System Design

Sweta Sharma, Soumen Santra, and Arpan Deyasi

2.1 INTRODUCTION

One of the most remarkable scientific advances in recent years has been the understanding that quantum information theory is a universal language for the behaviour of quantum systems [1]. Quantum effects are frequently a source of irritation for individuals who design current technologies.

Quantum tunnelling, for example, generates leakage current in transistors, quantum noise makes signals less certain, and quantum effects degrade the accuracy of time, space, and magnetic field measurements. According to quantum information processing theory and practice, these quantum effects are the advantages that can be used to manipulate the characteristics of electron transport in solid-state systems, achieve the quantum limits of communication and amplification, and go beyond the conventional limits to measurement [2].

2.1.1 Historical Development of Quantum Information Science

The universal theory of quantum mechanics governs the behaviour of matter at its most fundamental scales, which, therefore, governs the behaviour of information at those sizes. The most recent half-century of quantum research has revealed the Standard Quantum Limit, which controls the accuracy of clocks and interferometers in the absence of counterintuitive quantum effects such as squeezing and entanglement, and the Heisenberg limit, which establishes the ultimate quantum limits to metrology [3]. The quantum properties of light have revealed the ultimate potential of communication channels, including fibre optic communication channels and free-space communication via radio, microwaves, and light [4]. Because measurements inherently affect quantum systems, quantum cryptography is possible. Secret keys can be distributed using quantum communication channels, and their security is guaranteed by physical rules. Quantum physics governs computing power, as well as the number of bits that can be stored in physical systems and the speed at which those bits can flip.

DOI: 10.1201/9781003373117-2

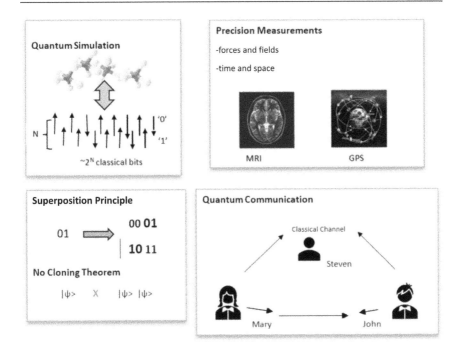

Figure 2.1 Quantum simulation for communication purpose.

The development of large-scale quantum computers and a quantum internet necessitates quantum error correction, quantum communication, and the development of coherent quantum interfaces between light and matter. The development of ever-larger-scale quantum computing has reached a tipping point. During the previous two decades, the density of quantum bits in general-purpose quantum computers did not quadruple every two years despite exponential advances in quantum technology. Figure 2.1 shows the quantum simulation output for communication purpose.

2.2 QUANTUM ALGORITHM

In quantum computing, the quantum algorithm is the algorithm used in the real model of quantum computing, and the most used model is the quantum chain model of computing. Although all classical algorithms can be implemented on a quantum computer, the term "quantum algorithm" is used for algorithms that appear quantum or use some important features of quantum computing.

Problems that cannot be solved using classical computers cannot be solved using quantum computers, as well. Shor's algorithm is faster than the most well-known classical algorithm for factorization, a common field of numbers. Grover's algorithm runs four times faster than the best classical algorithm for the same problem: linear search.

Quantum algorithms are usually described in terms of quantum circuits, which operate on multiple input qubits and end up measuring, in a widely used quantum computing circuit model. Quantum circuits consist of simple quantum gates that operate on a fixed number of qubits. The number of qubits must be fixed because the changing number of qubits leads to non-uniform evolution. Quantum algorithms can also be described in other quantum computing models, such as Hamilton's oracle model [5].

Quantum algorithms can be classified according to the basic techniques used by the algorithm. Techniques/ideas commonly used in quantum algorithms include phase impingement, phase estimation, quantum Fourier transform, quantum path, amplitude amplification, and topological quantum field theory [6]. Quantum algorithms can be classified according to the type of problem they solve; for example, see the FAQ on quantum algorithms for algebraic problems.

2.2.1 Algorithms Based on the Quantum Fourier Transform

The quantum Fourier transform is a quantum analogue of the discrete Fourier transform and is used in some quantum algorithms.

2.2.1.1 Deutsch-Jozsa Algorithm

The Deutsch-Jozsa algorithm solves the black-box problem, which may require too many black-box requirements for any deterministic classical computer, but can be done with a single query of a quantum computer. However, classical bounded-error and quantum algorithms lack speed in comparison, because classical probabilistic algorithms can solve problems with many consecutive queries with error probability [7].

2.2.1.2 Bernstein-Vazirani Algorithm

The Bernstein-Vazirani algorithm is the first quantum algorithm that solves problems more efficiently than the most popular classical algorithms [8].

2.2.1.3 Simon's Algorithm

Simon's algorithm solves black-box problems faster than classical algorithms, including finite error probability algorithms. This algorithm, which is exponentially faster than our classical computing algorithms [9], is the inspiration for Shor's factorization algorithm.

2.2.1.4 Quantum Phase Estimation Algorithm

The quantum phase estimation algorithm is used to determine the eigen-phase of the unitary gate eigenvector, which is proportional to the eigenvector and is given the quantum state for the gate input [10]. Algorithms are often used as subroutines in other algorithms.

2.2.1.5 Shor's Algorithm

Shor's algorithm solves discrete logarithm problems and integral factorization problems in polynomial time, while the most popular classical algorithms take super polynomial time. It is not clear whether this problem is complete in P or NP [11]. In addition, it is one of the few quantum algorithms that solve non-black-box problems in polynomial time, running in super polynomial time, among the most popular classical algorithms.

2.2.1.6 Hidden Subgroup Problem

A generalization of many problems that can be solved in a quantum computer, such as the Abelian hidden subgroup problem, the Simon problem, solving Pell's equation, checking the basic ideal of the ring R. There are known quantum algorithms for Abelian hidden subgroup problems. A more subtle group problem where the group is not necessarily abelian is a generalization of the previously mentioned problem and graph isomorphism and some lattice problems. Efficient quantum algorithms are known for non-abelian groups. However, there is no known efficient algorithm for symmetric groups that will provide an efficient algorithm for dihedral groups that will solve graph isomorphism and special compression problems.

2.2.1.7 Boson Sampling Problem

The Boson sampling problem in an experimental configuration involves averaging digital points (e.g., photons of light) to randomly scatter into many output modes bounded by known units. The problem is to generate a fair sample of the product probability distribution depending on the order of the input and the unitarity of the points. Solving this problem with classical computer algorithms requires constant calculation of the unitary transformation matrix, which may be impossible or prohibitively time-consuming. In 2014, it was proposed that existing technology and conventional probabilistic methods for generating single photon states could be used as inputs to quantum linear optical networks of computing, and quantum algorithms would provide a higher sampling of the output probability distribution. A 2015 study estimated that the sampling problem has similar complexity for inputs other than the Fock state photon and identified a difficult transition in computational complexity from the classical simulation problem to the Boson sampling problem, depending on the size of the inputs with uniform amplitude.

2.2.1.8 Estimating Gauss Sums

A Gaussian number is a form of exponential number. The well-known classical algorithm for evaluating this number takes exponential time. Quantum computers can approximate Gaussian quantities to polynomial accuracy in polynomial time.

2.2.1.9 Fourier Fishing and Fourier Checking

There is a word that consists of n random Boolean functions that map n-bit strings to Boolean values. For the Hadamard-Fourier transformation, we need to find n1-bit strings $z_1, \ldots, z_n$ that satisfy at least 3/4 of the string

$$\left| \tilde{f}(z_i) \right| \geq 1$$

and at least 1/4 satisfies

$$\left| \tilde{f}(z_i) \right| \geq 2$$

2.2.2 Algorithms Based on Amplitude Amplification

Amplitude amplification is a technique that allows the amplification of selected regions of the quantum state. Amplitude boosting applications typically yield a quadratic acceleration of the corresponding classical algorithm. This can be considered a generalization of Grover's algorithm.

2.2.2.1 Grover's Algorithm

Grover's algorithm searches an unstructured database (or unsorted list) with N entries for a given entry, using only $O\left(\sqrt{N}\right)$ queries instead of the $O(N)$ queries required classically. Classically, $O(N)$ queries are required even allowing bounded error probability algorithm.

Theorists considered a hypothetical generalization of a standard quantum computer that can access the history of hidden variables in Bohmian mechanics. Such a hypothetical computer could implement a search of an N-item database at most in $O\left(\sqrt[3]{N}\right)$ steps. This is quite faster than $O\left(\sqrt{N}\right)$ steps taken by Grover's algorithm. The search method will not allow any quantum computer model to solve NP-complete problems.

2.2.2.2 Quantum Computing

Quantum computing solves a generalization of the search problem. This solves the problem of counting the number of labelled entries in the unordered list instead of checking if there are any. More specifically, it counts the number of marked entries in an N-element list, with error ε making only $\theta\left(\frac{1}{\varepsilon}\sqrt{\frac{N}{k}}\right)$ queries, where k is the number of marked elements in a list. More precisely, the algorithm outputs an estimate k' for k, the number of marked entries, with the following accuracy: $|k - k'| \leq \varepsilon k$.

2.2.3 Algorithms Based on Quantum Walks

A quantum walk is a quantum analogue of a random walk, which can be represented by a probability distribution between certain states. Quantum tourism can be described by quantum superposition of states. Quantum excursions provide exponential speedups for some black-box problems. It also provides multi-threaded speed for many tasks. There is a framework for creating quantum computing algorithms, and it is a very versatile tool.

2.2.3.1 Element Distinctness Problem

The element difference problem is the problem of determining whether all elements of a list are different. Classically, the $\Omega(N)$ queries are required for lists of size N, but this problem can be solved in $\Omega\left(N^{2/3}\right)$ queries in quantum computers [12]. The optimal algorithm was proposed by Andris Ambainis. Yaoyun Shi first proved the low correlation when the range size is large enough.

2.2.3.2 Formula Evaluation

The problem is to evaluate the formula output from the root node given oracle access.

A well-studied formulation is a balanced binary tree with only NAND gates. This form of the formula requires $\theta(N^c)$ to use randomness where $c = log_2\left(1 + \sqrt{33}\right)/4 \approx 0.754$. With quantum algorithms, $\theta(N^{0.5})$ can be solved in terms of quantum algorithms that are better for this situation and are not known until unconventional Hamiltonian oracle models are found. For normal settings, the same result happens shortly.

For more complex formulas, fast quantum algorithms are also known.

2.2.3.3 Group Commutativity

The problem is to determine whether the black-box group given by generator k is commutative. A black-box array is a set of oracle functions that must be used to perform array operations (multiplication, inversion, and identity comparison). Deterministic and randomized query complexities are $\theta\left(k^2\right)$ and $\theta\left(k\right)$, respectively.

2.2.4 BQP-Complete Problems

The complexity class BQP (bounded error quantum polynomial time) is a set of solution problems that a quantum computer can solve in most cases, and in all cases, it has an error probability of at least 1/3 for all instances.

If there is a problem in BQP, it is full of BQP, and any problem in BQP can be reduced during polygamy. Informally, the full class of hard BQP problems is the hardest problem in BQP and can be solved efficiently (with bounded error) by quantum computers.

2.2.4.1 Quantum Simulation

The idea that quantum computers can be more powerful than classical computers comes from Richard Feynman's idea that classical computers require exponential time to simulate many-particle quantum systems. Since then, the idea that quantum computers can simulate quantum physical processes faster than classical computers has spread and developed. Efficient quantum algorithms (i.e., polynomial time) have been developed to simulate Bosonic and Fermionic systems and require several hundred qubits to simulate chemical reactions beyond the capabilities of current classical supercomputers. Quantum computers can also efficiently model topological quantum field theory. In addition to its intrinsic interest, this result leads to quantum algorithms for evaluating quantum topological transformations, such as Jones and HOMFLY polynomials, and Turaev-Viro invariant of three-dimensional manifolds. See Figure 2.2.

2.2.5 Hybrid Quantum/Classical Algorithms

A hybrid quantum/classical algorithm combines quantum state preparation with classical measurement and optimization. This algorithm usually aims to determine the eigenvector of the surface and the eigenvalue of the Hermitian Operator.

2.2.5.1 QAOA

The quantum approximation optimization algorithm is a quantum annealing game model that can be used to solve problems in the graph theory. The algorithm uses classical optimization of quantum operations to maximize the objective function.

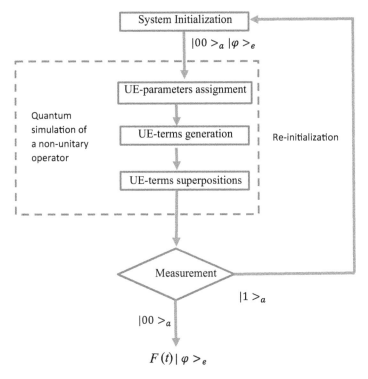

$$|00>_a |\varphi >_e$$

System Initialization

UE-parameters assignment

Quantum simulation of a non-unitary operator

UE-terms generation

Re-initialization

UE-terms superpositions

Measurement

$$|1>_a$$

$$|00>_a$$

$$F(t)|\varphi >_e$$

Figure 2.2 Flow diagram for simulating F(t) in a quantum computer. The system is initialized in the first block. The second block includes assignment of UE parameters, generation of UE terms, and superposition. Finally, measurements are made on the auxiliary subsystem to achieve a simulation of F(t) in a non-deterministic manner.

2.2.5.2 Contracted Quantum Eigensolver

The contracted quantum eigensolver (CQE) algorithm reduces the residual compression (or projection) of the Schrödinger equation into the space of two (or more) electrons to find the excited energy and the two-electron reduction of the density of the molecular matrix. This is based on the classical method to solve the energy and the two-electron reduced density matrix of the Schrödinger equation directly contracted against the Hermitian equation.

2.3 RECENT ADVANCES AND CURRENT PROSPECTS IN QUANTUM INFORMATION PROCESSING

Over the past ten years, there has been substantial progress in quantum computing, quantum communication, and quantum sensing and metrology.

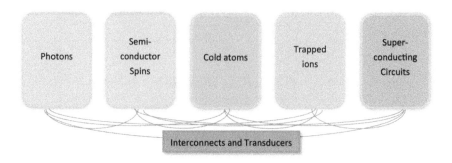

Figure 2.3 Block diagram connectivity of quantum activities

As previously stated, one of the driving causes behind this evolution has been the development of exponentially more precise fabrication, sensing, and control technologies—the same technologies that enable Moore's law for classical computer. Innovative quantum algorithms, more powerful error correction codes, novel tactics for quantum secure communications, and entanglement and squeezing-based approaches to quantum measurement are all key advances in the field of quantum information theory. The interplay of theory and experiment has sped up the development of quantum information processing systems. See Figure 2.3.

Some of the best physical platforms for processing quantum information are photons, spins in semiconductors, ultracold atoms, trapped ions, and superconducting circuits. Connecting these platforms is a critical field of research that is frequently required for the development of photonic transducers that can travel long distances with minimal decoherence.

2.3.1 Quantum Computation

Quantum computing is set to experience a dramatic transformation. Over the past several decades, quantum computation has been performed on relatively small-scale quantum systems of a dozen or more qubits. It has been shown that semiconductor quantum dots can be used in few-qubit systems. A topological system-based approach to processing quantum information is extremely promising but difficult to implement [13].

Quantum dots, semiconductor systems that mimic atoms, and semiconductor dopants all can perform multiple quantum logic operations on many qubits. Due to lithographic manufacturing procedures, like those used to produce conventional silicon-based microprocessor chips, semiconductor quantum information processors may offer a high level of scalability and architectural freedom.

Until now, however, these simulations could be done on very large, and occasionally specialized, classical computers. The threshold for quantum computers to supplant all other tools as the most powerful means of characterizing quantum behaviour in a wide range of naturally occurring quantum systems is approaching. At the scale at which classical computation becomes impossible, quantum computers can begin modelling the behaviour of other quantum systems with the precision that conventional computers cannot reach [14]. The most powerful near-term use of quantum computation is to simulate condensed matter and particle physics models, including quantum gravity models. Over the past ten years, quantum simulation demonstrations have improved rapidly, thanks in part to the development of a number of cutting-edge methodologies for modelling quantum chemistry.

Large-scale quantum annealers built with superconducting circuits enable the modelling of customizable transverse device for solid-state quantum information processing. Entanglement can be utilized to create semiconductor quantum dots, atomic spins in solids, and extraordinarily sensitive detectors of electric and magnetic forces utilizing nitrogen vacancies in diamond. Special-purpose quantum information processors have already enabled the simulation of enormously complex quantum systems whose behaviour is much beyond the capabilities of any classical computer.

2.3.2 Theory of Quantum Computation

Despite the paucity of new quantum algorithms, quantum computer research has made enormous advances in the recent decade, exploring large areas previously unexplored. The concept of Quantum Merlin Arthur (QMA) completeness, which is the quantum analogue of NP-completeness, has prompted various investigations into whether quantum computers may be built by manipulating their Hamiltonian dynamics. Recent research into the power of simple quantum circuits of the type that experiment is likely to make available in the coming years has been motivated by the concept of "quantum supremacy," which holds that even simple quantum systems can generate data with the statistics that no classical computer can generate [15].

The feasibility of developing large-scale quantum computers has been considerably influenced by advances in error correction coding theory over the past two decades. Quantum error correction is required for devices that can factor huge numbers to perform reliably. Building such large-scale quantum computers remains a massive endeavour that will take at least twenty years to complete, but developments in quantum error correction code thresholds during the past ten years have shown that it is at least theoretically conceivable.

2.3.2.1 Three-Bit Code

We begin with a detailed analysis of the operation of the simplest quantum error correction code. Let us say that a source A sends quantum information to the receiver through a noisy communication channel B. Of course, the channel must be noisy in practice, because no channel is too quiet. However,

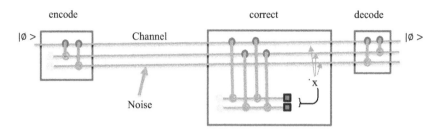

Figure 2.4 A simple example illustrating the principle of quantum error correction. Mary wanted to transmit one-qubit state $|\varnothing> = a|0> + b|1>$ to John through the input channel that introduces σ_x errors ($|0> \leftrightarrow |1>$) randomly. Mary prepares two more qubits in the state $|0>$ indicated by the small circle. These three qubits are sent to John. At the receiving end, John restores the joint state by outputting its syndromes and the corrections based on these syndromes. The correction is the σ_x function used by one (or none) of the qubits. Finally, the decoding operation separates one qubit from the others and gives John a single qubit in the state $|\varnothing>$ with probability $1 - O(p^2)$.

to do better than just sending quantum bits down the channel, you need to know about its sound. The following properties are considered for this input section: sound effects per qubit are independent, and for a given qubit, there is a randomly chosen effect between exiting the qubit, constant state (probability 1—p), and using the Pauli operator σ_x (probability $p < 1/2$). This is a very artificial sound, but if we fix it, we will see what fix we can offer for useful results that lead to a more realistic sound type.

The simplest quantum error correction method is summarized in Figure 2.4.

2.3.3 Quantum Communication

During the past ten years, all three of these quantum communication applications have achieved significant progress. The theoretical limit of these channels has finally been established in the past two years; while we still don't know how to reach it in actuality, we know what the goal is. According to the concept of distributed quantum computation over a quantum internet, linked quantum computers could enable users to exchange joint information in a trustworthy manner, ranging from quantum secret sharing and quantum decision making to fraud-free quantum voting. The quantum internet has the potential to change how individuals and organizations interact and compete, increasing trust while maintaining privacy, just as the traditional classical internet has enabled a wide range of applications that go beyond the use of a single classical computer [16].

The field of quantum communication has several applications and far-reaching consequences. Quantum mechanics enables the development of physically secure communication protocols, such as quantum data locking, quantum secret sharing, and quantum key distribution.

Quantum data locking and "blind" quantum computation, which use quantum mechanics laws to preserve stored information and render a quantum computer inaccessible to outside operators, provide innovative solutions to computer security. They make quantum computers linked in a quantum internet far less resistant to attack than traditional computers [14].

2.3.4 Quantum Sensing and Metrology

Reaching those boundaries necessitates paying attention to how "quantum weirdness" manifests itself in the context of sensing and measurement. Quantum information theory, for example, has demonstrated that the ultimate limits of interferometry can be attained by infusing exotic quantum states, such as compressed vacuum states, into interferometer ports.

The development of quantum-limited vibrational sensors and accelerometers over the past few decades has marked a significant progress in precision quantum measurement. Mechanical cantilevers made of nanoscale materials can be created and instrumented to perform at the standard quantum limit [17]. Individual quantum cantilevers can operate at the standard quantum limit, but theoretical proposals to entangle such cantilevers with light and with one another raise the prospect of developing nanomechanical systems that can operate at the Heisenberg limit for metrology and could be used to detect quantum gravity effects.

Ion trap quantum information processors can be used as entangled quantum clocks to attain an exceptional temporal precision. The applications and uses of entanglement for precision time measurement are described in the quantum information theory. As previously stated, precise optical frequency clocks are dependent on entanglement in the quantum logic atomic clock. By constructing a linked quantum network of atomic clocks, the benefits of squeezing and entanglement in measurements could be extended to a quantum sensor network.

The resulting network may use entanglement to provide safe and precise "quantum GPS," for example, by measuring position and acceleration very precisely. Optical frequency atomic clocks are already sensitive gravitational field detectors due to gravitational redshift. Extensive networks of entangled quantum clocks could serve as exceedingly accurate gravitometers to investigate the structure of seismic faults.

The construction and application of precise, coherent quantum transducers lie at the heart of quantum sensing. The amazing success of existing quantum sensors and measurement equipment can be attributed to the quantum technologies that enable the fabrication of quantum transducers. The development of more efficient quantum interfaces and transducers will be critical in the future for the development of ever-more accurate and precise quantum sensors [18].

2.4 CONCEPTUAL AND TECHNICAL CHALLENGES

To fully realize the potential of quantum information processing technology, we must overcome a number of challenges. Theory and experiment interact intimately in quantum information processing, implying that conceptual and technological issues are tightly intertwined. The main challenges confronting each of the three subfields of quantum information processing are covered in this section. These obstacles are intertwined in each sector.

2.4.1 Challenges for Quantum Computing

Numerous technical challenges must be addressed to produce reliable mid-scale quantum computers, in terms of both quantum scalability and design and build concerns. Research is essential to understand and eliminate the causes of noise and mistakes. To reduce noise in ion traps and superconducting quantum circuits, fresh fabrication and design methods, as well as novel quantum control methods, are required.

Even though ion-trap and superconducting quantum computers are currently the most plausible platforms for constructing mid-scale quantum computers, it is still too early to focus solely on basic research in these technologies. Systems based on semiconductors and atoms provide exceptional scalability and coherence [19]. Topological systems are also promising due to their natural resilience to noise and errors. Even though topological techniques of quantum information processing are still in their infancy, advances in system architecture and material fabrication may make these methods great candidates for creating mid- and large-scale quantum computers [20].

Individual qubits should be designed to naturally resist decoherence and errors, couplings between qubits should be designed to allow high-fidelity quantum logic gates while minimizing crosstalk, input-output ports should be implemented in a way that allows high-accuracy preparation and measurement, and the coupling of quantum information with quantum communication should all be considered during the design phase of any experimental quantum logic implementation.

The second line of defence employs techniques such as dynamical decoupling and more advanced quantum management algorithms to reduce noise and mistakes at the single-qubit level and assure high-accuracy quantum logic operations. The third line of defence against errors and noise is quantum error correction codes. See Figure 2.5.

Creating fault-tolerant quantum computation necessitates knowledge of ever-more advanced technology. Quantum algorithms and quantum no destruction measurements can now be performed on a few qubits. Logical qubit encoding with a performance superior to physical constituent qubits is projected to be achieved within the next five years.

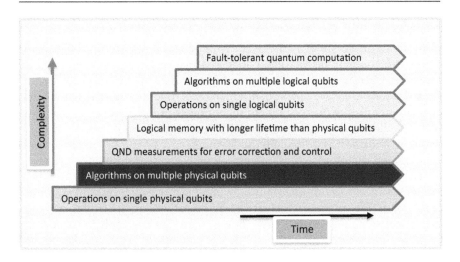

Figure 2.5 Quantum technologies framework

2.4.1.1 Quantum Error Correction

Over the past ten years, the theoretical side of the discipline has made the most progress in quantum error correction, improving on existing coding schemes and establishing new ones, such as surface codes, with higher error correction thresholds. One of the most pressing concerns confronting quantum information theory is how to continue development in generating stronger quantum error correction algorithms [21].

The application of topological systems to fault-tolerant quantum computation is a promising technique. Topological systems are inherently immune to quantum mistakes: local faults have no effect on computation. The design and construction of fault-tolerant topological quantum computers is a hybrid theory/experiment challenge of effective two-dimensional topological systems. Despite tremendous progress in identifying materials and systems with the required topological features, much more work remains to be done if topological quantum computation is to become a reality.

2.4.1.2 Algorithms for Mid-Scale Quantum Computers

Mid-scale quantum computers will be able to demonstrate quantum supremacy by producing outputs with probability that classical computers cannot. Near-term demonstrations of quantum supremacy, however, do not always have practical implications (other than demonstrating that quantum computers can perform things that classical computers cannot). The creation of novel and usable types of quantum supremacy is a significant problem for mid-scale quantum computation.

2.4.1.3 Special-Purpose Quantum Information Processors

Quantum annealers will soon have a wide range of applications. They can be used to simulate fascinating, condensed matter physics models, such as a number of transverse field Ising models and their quantum phase transitions. Because they are open quantum systems functioning at nonzero temperature, they allow us to simulate such models in the presence of coupling with a well-characterized environment. One of the most promising uses of quantum annealers is deep quantum learning.

Integrated quantum optical systems, which are dense, tunable interferometric arrays etched into a silicon chip, are one of the most astounding recent advancements in quantum information processing technology. Integrated photonic systems, which are based on technology for building classical optical switching arrays, have shown to be effective tools for executing quantum information processing. When paired with effective photon-counting detectors and nonclassical light sources, integrated linear optical systems represent a potentially scalable alternative approach to the building of universal quantum computers.

2.4.2 Challenges for Quantum Communications

A quantum internet promises long-distance secure communication, precision sensing and navigation, and distributed quantum information processing, which are all unattainable in a classical environment. A quantum internet will most likely consist of several types of quantum memories and quantum information processors connected by optical lines. As previously said, it will incorporate many distinct forms of quantum memory and sensors, such as cold atoms, solid state, photonic, and superconducting. This heterogeneity parallels an essential component of today's "classical internet of things," as it connects diverse information processing systems, each with its own set of benefits and drawbacks.

2.4.2.1 Quantum Key Distribution (QKD)

Recently, theoretical constraints on the rate at which secret keys can be created as a function of communication distance were demonstrated. When comparing today's experimental demonstrations to theoretical rate limits, improved experimental hardware and protocol implementations can still produce a two-order-of-magnitude gain at a given loss. Even with the greatest procedures, after a few hundred kilometres, the rate will approach zero as photon transmission virtually vanishes.

Almost all encrypted internet traffic nowadays employs asymmetric public-key cryptography. Asymmetric encryption systems, on the contrary, such as the commonly used RSA algorithm [22], are not inherently secure; their security is reliant on the unproven premise that an eavesdropper lacks the computational resources or procedures required to decrypt ciphertext. Asymmetric encryption techniques, such as the one-time pad, however, guarantee absolute secrecy but require users to hold secret, identical keys. Quantum physics properties are used to amplify secret information at a distance in the only provably secure way.

2.4.2.2 Quantum Repeater Networks/The Quantum Internet

Small quantum networks on the metro scale are being actively researched and developed. The first entangled quantum memories between partial repeater nodes have been demonstrated, but considerable obstacles remain. The QKD field deployments stated earlier are crucial in the development of this precise network technology. Additional qubits are required for quantum error correction, entanglement distillation, and quantum repeater multiplexing. Auxiliary components such as quantum frequency conversion and on-demand entangled photon-pair generators enable the network to enhance entanglement rates, extend link distances, and enable fault tolerance. Both components have made great progress, but further efficiency and scalability improvements are required.

Improved quantum repeater protocols appear to have a great deal of promise for lowering the technological obstacles to high-speed quantum repeater networks. Protocols that use feed-forward error correction, which eliminates most of the time-consuming, two-way classical communication between repeater nodes, are particularly promising. Error-corrected, loss-protected photonic entangled states, as well as small-scale photonic quantum logic processors, are required for such protocols. One goal would be to create loss-protected optical quantum channels capable of retaining encoded photonic quantum states in the laboratory and across deployed optical channels for a longer period than low-loss optical fibre. In this

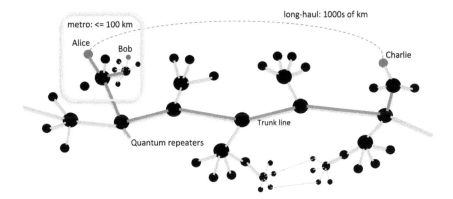

Figure 2.6 Quantum key distribution (QKD) diagram

instance, improved on-demand photon-pair generators and detectors are also essential.

Connecting distant quantum memories will be the next step in quantum networks. In a basic repeater network, atomic quantum memories along a connection are pairwise entangled and coupled via free space or fibre links. See Figure 2.6.

2.4.3 Challenges for Quantum Sensing and Metrology

Because interferometers, atomic clocks, magnetometers, gyroscopes, and accelerometers are already working at the standard quantum limit, quantum sensing and metrology are the most mature and widely utilized applications of quantum information processing technology. The main issue for quantum sensing and metrology is to go beyond the standard quantum limit to achieve the ultimate, Heisenberg-limited precision and accuracy constraints [15].

The combination of theoretical and experimental problems that must be addressed to achieve the ultimate quantum limits of precision measurement is emphasized.

2.4.3.1 Interferometry

Optical interferometers are among the oldest precision quantum technologies, and they were among the first to achieve the standard quantum limit.

The construction of integrated quantum optical chips containing hundreds of tunable interferometers in a square centimetre is one of the promising advances in quantum information processing technologies described earlier. The ultimate quantum precision that such devices provide for measuring changes in distance remains an open question.

2.4.3.2 Quantum Clocks

The fundamental difficulty for showing more accurate and precise quantum clocks is to connect precision quantum measurement theory with the practice of manufacturing quantum clocks that leverage quantum processes such as entanglement to maintain the world's most precise measurement devices. As will be demonstrated later, by connecting such quantum clocks via quantum communication channels to establish a global quantum GPS network, global timekeeping precision can be boosted by the orders of magnitude.

2.4.3.3 NV-Diamond Sensors

Nitrogen vacancy diamond (NV-diamond) combines in a single system (an NV-center) the precision control and readout of atom-optical systems with the robustness of solid-state systems. Nitrogen vacancies behave like "super atoms" embedded in a diamond lattice. Such states have the potential to provide highly precise measurements of the electromagnetic field at the nanoscale level. The current use of NV-centers for magnetometry and electric field sensing operates largely at the semi-classical standard quantum limit.

They can also function as precise thermometers to measure the temperature in their local nanoscale environment. Spin-orbit couplings allow NV-centers to function as single-atom gyroscopes to measure changes in rotation and orientation. Finally, if coupled with photons, NV-centers have the potential for modular universal quantum computation and quantum repeaters.

Their ubiquitous utility renders NV-centers one of the most promising substrates for quantum sensing, imaging, and information processing. Much work remains to be done, however. The current quantum applications of NV-centers arose out of advances in fabrication technology for NV-diamond, by allowing the relatively precise implantation of NV-centers in an ultra-pure and isotopically enhanced diamond. An important challenge to the development of NV-diamond quantum technologies is the ability to implant NV-centers with atomic precision. If this were possible, crystalline arrays of NV-centers could allow improved error-corrected logical qubits or could be used for room-temperature solid-state quantum computation.

Nitrogen vacancies are not the only type of atom-like defect that is available for quantum information processing. Silicon vacancy centres are significantly more stable optically, and there is a promise that they, too, will feature long spin coherence times. SiC di-vacancies, though less studied than NV-diamond, represent a highly promising system for quantum information processing, including metrology and computation. A key challenge over the next twenty years is the ongoing development of material and fabrication techniques to put to use controllable, atom-like systems in the solid state.

2.4.3.4 Nanomechanical Resonators

A fundamental characteristic of matter is the quantification of vibrational degrees of freedom, known as phonons. Recent advances in quantized mechanical resonating devices, such as nanomechanical cantilevers, show a considerable promise for sensing acceleration and basic gravitational phenomena. This type of squeezed and entangled nanomechanical resonator could be utilized as a probe to measure acceleration and gravitational fields beyond the standard quantum limit. Because they are heavier than individual atoms and electrons, quantized vibrational mechanical systems can give sensitive testing of the predictions of many quantum gravity theories.

The ability to connect quantum computers to a quantum internet via quantum communication lines opens a slew of potential applications for extended quantum computation, just as the ability to connect quantum sensors and measurement devices via quantum communication channels opens a slew of potential applications for quantum sensing and imaging.

2.5 CONCLUSION

The US government's support for basic research in quantum information processing during the past two decades has sown the seeds of a whole new field of science and industry. Continued research into the fundamental physics of quantum information processing over the next two decades will allow us to reap the quantum crop. This technical advancement was enabled and inspired by the creation of a comprehensive theory of quantum information. Because of the growing speed of technological and theoretical growth, the field of quantum information has reached a critical juncture. At the moment, there is a clear path to developing medium-sized quantum computers that can solve problems that classical computers cannot; quantum cryptography is real, and the development of quantum repeaters holds the promise of global quantum-secured communications and a slew of new applications; and the possibility of building measurement devices that achieve the Heisenberg limit has been demonstrated, and methods for doing so have been developed. Similarly, new channel encoding techniques are necessary to reach quantum channel capacities.

REFERENCES

[1] I. Y. Aref'eva, I. V. Volovich, "On the large time behavior of quantum systems", Infinite Dimensional Analysis, Quantum Probability and Related Topics, Vol. 4, No. 4, pp. 453–482, 2000

[2] A. Hobson, "Entanglement and the measurement problem", Quantum Engineering, Vol. 2022, A. ID. 5889159, 2022

[3] B. Schumacher, "Quantum coding", Physical Review A, Vol. 51, No. 4, pp. 2738–2747, 1995

[4] A. Beveratos, R. Brouri, T. Gacoin, A. Villing, J.- P. Poizat, P. Grangier, "Single photon quantum cryptography", Physical Review Letters, Vol. 89, p. 187901, 2002

[5] A. Montanaro, "Quantum algorithms: An overview", npj Quantum Information, Vol. 2, No. 15023, 2016

[6] A. Fowler, M. Mariantoni, J. Martinis, A. Cleland, "Surface codes: Towards practical large-scale quantum computation", Physical Review A, Vol. 86, p. 032324, 2012

[7] D. Collins, K. W. Kim, W. C. Holton, "Deutsch-Jozsa algorithm as a test of quantum computation", Physical Review A, Vol. 58, p. R1633, 1998

[8] A. Shukla, P. Vedula, "A generalization of Bernstein–Vazirani algorithm with multiple secret keys and a probabilistic oracle", Quantum Information Processing, Vol. 22, No. 244, 2023

[9] R. J. Lipton, K. W. Regan, "Simon's algorithm", Quantum Algorithms via Linear Algebra: A Primer, MIT Press, pp. 93–96, 2014

[10] O. Ouedrhiri, O. Banouar, S. E. Hadaj, S. Raghay, "Quantum phase estimation based algorithms for machine learning", IEEE 2nd International Informatics and Software Engineering Conference, 2021, Ankara, Turkey

[11] P. W. Shor, "Polynomial-time algorithms for prime factorization and discrete logarithms on a quantum computer", SIAM Journal on Computing, Vol. 26, No. 5, p. S0097539795293172, 1997

[12] H. Buhrman, C. Durr, M. Heiligman, P. Hoyer, F. Magniez, M. Santha, M. de Wolf, "Quantum algorithms for element distinctness", 16th Annual IEEE Conference on Computational Complexity, 2001, Chicago, IL, USA

[13] V. Lahtinen, J. K. Pachos, "Topological aspects of quantum information processing", Topology and Condensed Matter Physics, Vol. 19, chapter 18, Springer, pp. 471–500, 2017

[14] X. Waintal, "What determines the ultimate precision of a quantum computer", Physical Review A, Vol. 99, p. 042318, 2019

[15] A. M. Childs, J.-P. Liu, A. Ostrander, "High-precision quantum algorithms for partial differential equations", Quantum, Vol. 5, p. 574, 2021

[16] N. Gisin, R. T. Thew, "Quantum communication technology", Electronics Letters, Vol. 46, No. 14, pp. 965–967, 2010

[17] P. Sekatski, M. Skotiniotis, J. Kołodyński, W. Dür, "Quantum metrology with full and fast quantum control", Quantum, Vol. 1, p. 27, 2017

[18] N. Aslam, H. Zhou, E. K. Urbach, M. J. Turner, R. L. Walsworth, M. D. Lukin, H. Park, "Quantum sensors for biomedical applications", Nature Reviews Physics, Vol. 5, pp. 157–159, 2023

[19] Z. Xi, Y. Li, H. Fan, "Quantum coherence and correlations in quantum system", Scientific Reports, Vol. 5, No. 10922, 2015

[20] A. Erhard, J. J. Wallman, L. Postler, M. Meth, R. Stricker, E. A. Martinez, P. Schindler, T. Monz, J. Emerson, R. Blatt, "Characterizing large-scale quantum computers via cycle benchmarking", Nature Communications, Vol. 10, No. 5347, 2019

[21] J. Roffe, "Quantum error correction: An introductory guide", Contemporary Physics, Vol. 60, No. 3, pp. 226–245, 2019

[22] X. Zheng, Z. Zhao, "Quantum key distribution with two-way authentication", Optical and Quantum Electronics, Vol. 53, No. 304, 2021

Chapter 3

Automatic Classification of Tables Using Hybrid Quantum Convolutional Neural Networks

Erick Franco-Gaona, Iván Cruz-Aceves,
and Maria-Susana Avila-Garcia

3.1 INTRODUCTION

Research documents often include information elements such as figures, equations, and tables that provide important information that needs to be searched, accessed, and retrieved. However, extracting and structuring the information contained in these elements is not an easy task. Figures, equations, and tables can be considered unstructured information elements (UIEs), given that these cannot be processed and analyzed using conventional methods. More sophisticated solutions are needed to automate the location, access, classification, and data extraction from them. Some of these methods can include computer vision techniques, machine learning algorithms, and deep learning solutions.

For instance, for mathematical expression detection, a two-stage hybrid method was proposed by Phong et al. [1]. Deep learning and hand-crafted capabilities are combined to improve detection accuracy using AlexNet and ResNet architectures as the basis of pre-trained networks. The proposed method has been evaluated on two datasets, Marmot dataset for equations [2] and GTDB dataset [3], obtaining better results with Marmot and a competitive performance in the literature. Lin et al. [4] suggested a technique that merges rule-based and learning-based approaches to recognize individual mathematical expressions in PDF documents. They also employed different characteristics of the formulas, such as geometric structure, character, and context contents, to make it suitable for recognizing a broad variety of formula types. The experimental results demonstrate that the suggested method performs satisfactorily.

The deep learning technique presented by Kukkadapu et al. [5] can detect formulas in a document page image by sliding windows at different scales. The method clusters candidate detections to produce page-level outcomes. The study utilized a modified version of the GTDB scanned mathematical article collection and demonstrated promising results but with possible enhancements in the future.

On the contrary, Luo et al. [6] introduced a technique for extracting data from diverse kinds of graphs. The approach integrated a deep framework

DOI: 10.1201/9781003373117-3

and rule-based approaches to achieve adequate generalization capacity and to acquire precise interim results. The experiments demonstrated that the method attained high performance with rapid processing speed when trained on the 400KExcelcharts dataset for deep model training. Liu et al. [7] conducted a study on different table formats to identify table headers by randomly selecting data from the CiteSeerX dataset. The results of their analysis of PDF documents indicate that utilizing a Random Forest classifier yields comparable accuracy to that reported in existing literature.

This work focuses specifically on the classification problem for tables, identifying whether an image provided as an input is a table or not.

Convolutional neural networks (CNNs) were applied on image classification problems in the past [8–12]. CNNs are defined so that each component is trained to perform a task, which reduces the number of hidden layers and training. CNNs can detect from simple features, such as edges, to more complex features, such as detecting a target [13]. However, training of a CNN takes considerable time and resources. Transfer learning (TL) is a solution for this issue. TL allows to use the knowledge obtained by a trained network in one scenario as a solution for another (similar) problem, thus avoiding training the network from the scratch [14]. Including TL implies adjusting the CNN hyperparameters. Hyperparameters are values utilized to regulate the function and the intricacy of a neural network [13]. A few prevalent illustrations of hyperparameters in a neural network include the quantity of hidden layers, the rate of learning, the quantity of convolution filters, and so on. Some researchers have used TL with previously trained networks to extract unstructured information elements.

For instance, to detect tables in digital documents, Schreiber et al. [15] used the ICDAR 2013 table competition dataset [16] and the Marmot table dataset to recognize tables and their structures with VGG-16 architecture. They obtained an F1-score of 96.77% and 91.44% in table and structure detection, respectively. Paliwal et al. [17] proposed a VGG-19 architecture as the basis of the neural network for table detection. Furthermore, the authors tagged elements in the Marmot database using boxes to label the location of columns and rows for detecting table structure. The results showed that the use of semantic features further improved the model performance by leveraging TL and obtained F1 Scores for table detection and table structure recognition of 95.47% and 92.15%, respectively.

DeCNT, the approach presented by Siddiqui et al. [8], merges faster R-CNN/FPN with a deformable convolutional neural network. Deformable convolution can modify its receptive field to match the input, allowing it to adapt to the input. Through this technique, the researchers were able to attain an F1-score of 89.5%. The neural network presented by Agarwal et al. [9] incorporates a dual backbone architecture that also uses deformable convolution to detect tables of varying sizes. The authors performed an empirical assessment of their CDeC-Net model achieving a maximum F1-score of 95.2% on the Marmot dataset.

Nazir et al. [10] introduced HybridTabNet, which identifies tables in scanned document images. The table detector leverages ResNeXt-101 as a backbone. Furthermore, the backbone integrates deformable convolutions. The researchers achieved a maximum F1-score of 95.6%. Hashmi et al. [11] employed Graph Neural Networks (GNN) to detect tables in digital documents using a lightweight ResNet-50 backbone. They were able to attain a maximum F1-score of 95.8%. Kazdar et al. [12] suggested that DCTable is an approach to enhance faster R-CNN for detecting tables. DCTable utilizes a backbone with dilated convolutions to extract more distinctive features, which leads to better region proposals. The maximum F1-score obtained was 96.9%.

Currently, quantum machine learning is a new research area, related to quantum mechanics and computer science. Quantum machine learning models use the advantages of quantum information to improve classical machine learning [18]. One example of using quantum computing is through quantum convolutional neural networks (QCNNs). Kwak et al. [19] defined a QCNN as a type of neural network that incorporates the concepts of quantum computing and CNNs. Instead of using classical bits, QCNNs use qubits as the fundamental unit of information processing. This feature enables them to execute specific tasks more proficiently than classical CNNs. One of the main benefits of QCNNs is their ability to process information simultaneously by utilizing the principles of quantum superposition and entanglement, whereas classical CNNs process information layer by layer. In addition, QCNNs can handle significant datasets more efficiently than classical CNNs.

Another approach is the use of a hybrid quantum convolutional quantum neural network (HQCNN), which combines the features of a QCNN with a traditional CNN that applies hybrid transfer learning (HTL) to enhance classification accuracy. In this case, HTL refers to quantum-to-classical transfer learning. A pre-trained quantum system behaves as a kind of feature extractor. Subsequently, a classical network is used to process the extracted features for the specific classification problem [20]. The implementation of HQCNNs has been reported in image classification problems. For instance, for medical images, Houssein et al. [21] used an HQCNN to detect Covid-19 in X-ray images, obtaining better results than the literature. Kulkarni et al. [22], using a classical network for feature extraction and a quantum circuit to classify, obtained better results than with traditional methods to classify pneumonia from chest radiographs.

Using an HQCNN for detecting stenosis in X-ray coronary angiography, Ovalle-Magallanes et al. [23] demonstrated that their hybrid proposal is more effective than non-quantum methods. On the contrary, Sebastianelli et al. [24] described the classification of surfaces from satellite images using an HQCNN having a higher performance than with classical networks. Despite the potential of HQCNNs, their implementation to classify UIEs, such as tables, has not been reported in the literature yet.

In this chapter, an HQCNN is proposed for the classification of tables in digital documents.

3.2 BACKGROUND

This section presents in detail the fundamentals of the techniques used to classify information elements, specifically CNN and QCNN. In addition, the data set with which the tests are carried out and how it has been used in other works is introduced.

3.2.1 Database of Information Elements

In this work, the Marmot dataset is used for table detection and classification [2]. Marmot dataset contains 2,000 PNG images extracted from PDF documents that are collected from the following:

- The Founder Apabi digital library, which contains PDF research documents in Chinese. A total of 120 e-books from different subjects were extracted.
- Around 1,500 conference and journal articles published in English and Chinese obtained from 1970 to 2011.

Images in the dataset are tagged as either table or non-table as shown in Figure 3.1. The dataset includes different types of tables such as ruled, partially ruled, and unruled, as well as horizontal, vertical, internal column, and split column tables; examples of these types of tables are shown in Figure 3.2. The dataset includes both "positive" and "negative" instances, representing the presence and absence of tables, respectively, in an equal 1:1 proportion. There are 1,000 pages that have at least one table, while the other 1,000

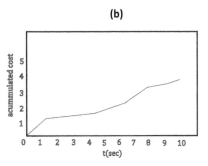

(a)			
Scheduling	J1	J2	J3
RM	0.0033	0.0930	8
EDF	0.0033	0.0930	0.1769
LEF	0.0033	0.0812	0.1645

Figure 3.1 Examples of (a) table and (b) non-table. (Recreated from Marmot dataset)

(a)

Scheduling	J1	J2	J3
RM	0.0033	0.0930	8
EDF	0.0033	0.0930	0.1769
LEF	0.0033	0.0812	0.1645

(b)

Scheduling	J1	J2	J3
RM	0.0033	0.0930	8
EDF	0.0033	0.0930	0.1769
LEF	0.0033	0.0812	0.1645

(c)

Scheduling	J1	J2	J3
RM	0.0033	0.0930	8
EDF	0.0033	0.0930	0.1769
LEF	0.0033	0.0812	0.1645

(d)

Scheduling	RM	EDF	LEF
J1	0.0033	0.0033	0.0033
J2	0.0930	0.0930	0.0812
J3	8	0.1769	0.1645

Figure 3.2 Examples of (a) ruled table, (b) partially ruled table, (c) unruled table, and (d) vertical table. (Recreated from Marmot dataset)

pages do not have tables but may have page components that resemble tables, such as arrays and figures.

In this work, a new dataset of table images was created from the Marmot dataset. Table images were extracted, and individual images were created to classify them as either table or non-table. As shown in Figure 3.3, the document contains tables and more UIEs, so it is necessary to individually separate the tables.

3.2.2 Convolutional Neural Networks

CNNs are often used in object classification and detection problems in both the images and the videos in real time. CNNs, such as basic neural networks, consist of layers; however, as illustrated in Figure 3.4, certain layers in CNNs perform specific functions such as convolution and pooling for features' extraction and fully connected layers for classification [25]. A CNN is trained to classify an input image by identifying different levels of abstract feature representations, ranging from basic features such as edges to more intricate ones such as specific parts of an object. The types of layers in a CNN can be briefly defined as follows:

- Convolution layer: A layer that applies a convolution operation to the input data by means of filters generating feature maps.

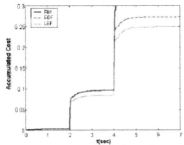

Fig. 2. Accumulated cost under different policies.

with the difference in performance due to the use of different scheduling policies, which is the objective of our simulations.

The performance of each pendulum is obtained by the following cost function J_i, which measures the error e_i of the pendulum weighted with time t.

$$J_i(t) = \int_0^t t e_i^2(t) dt$$

Note that t sets the duration of the evaluation period, which typically should go from the perturbation arrival time (0 in the integral) to the settling time. As higher values the cost function gets, the worst the control performance (because major deviations occur or because it takes more time for the inverted pendulum to recover from the perturbations).

B. Results

The simulation we run keeps the following time sequence: at time t=0, only the control task controlling the fist pendulum is released. After executing alone, at time t=2, another control task is released to control the second pendulum. Finally the third control task is released at time t=4. The three controllers run in parallel until t=7. Before the release time of each control task, the corresponding pendulum is in equilibrium. At release time of each control task, the pendulum suffers a perturbation (of equal magnitude for each pendulum). This simulation pattern is repeated for different scheduling algorithm: RM, EDF and LEF. During each experiment, the cost function presented earlier and the resulting schedule are recorded. The cost function evaluation period goes from the beginning of each simulation, t=0, to the simulation completion, at t=7. For each scheduling policy, the results we obtained are summarized in the following:

- **RM:** RM is a static scheduling algorithm in open loop, which assigns priorities to tasks according to their request rates. The cost function values for the three pendulums are shown in the Figure 2 (note that Figure 2 shows the accumulated cost function, ΣJ_i, during the evaluation period for each scheduling policy). Under RM the schedulability condition for our experiment is given by U < 0.78. Taking into account that task execution time is 4ms, until t=4, the processor utilization is 0.71 and the control performance is good. From t=4, the tree control tasks run in parallel and

these consume 0.99 CPU. That is, task3 misses deadlines and the inverted pendulum1 falls down. This explains the fact that the cost function in Figure 2 goes to infinite. Note that under RM, the task is not schedulable.

- **EDF:** EDF is a dynamic algorithm in open loop, which assigns priorities to tasks according to their absolute deadlines. Under this scheduler the schedulability condition is given by U < 1. For our simulations, since U < 1, the task set is schedulable and the three pendulums can be controlled as it can be seen in Figure 2: the accumulated cost reaches a finite value, which means that the deviation caused by each perturbation that affected each of the three pendulums could be adequately corrected. The performance achieved by EDF is also given in Figure 2 in terms of the cost function, reaching a value of 0.2732 at the completion of the evaluation interval

- **LEF:** LEF is a dynamic scheduling algorithm in closed-loop, which adjusts the schedule based on continuous feedback of each control loop. Under this scheduler, in the simulation we obtain that the three inverted pendulums can be perfectly controlled. As it can be seen in Figure 2, the cost function of LEF goes below the cost function of EDF, meaning that for this particular simple simulation set-up, LEF performs better that EDF. Note that the final cost of LEF is 0.2490.

C. Discussion

The exact cost for each one of the three pendulums under each scheduling algorithm is summarized in the Table 1. RM fails in controlling the third pendulum. EDF and LEF are able to control the three pendulums. However LEF gives the best control in terms of the cost function.

Table 1. Cost for the three inverted pendulums under each different scheduling policy.

Scheduling	J_1	J_2	J_3
RM	0.0033	0.0930	8
EDF	0.0033	0.0930	0.1769
LEF	0.0033	0.0812	0.1645

This results shows that scheduling policies that take advantage of the application dynamics and are able to adjust the schedule accordingly can provide, in some specific scenarios, better performance in terms of the application (control performance in our case).

For open-loop scheduling approaches we identified (see Section 1) three main negative aspects: low real CPU utilization, the lack of feedback mechanism and poor adaptability to the application dynamics. Our strategy optimizes CPU utilization in the sense that control tasks are executed only when they are required. That is, they are executed when the controlled plants suffer perturbations. Otherwise, they execute with the slowest possible rate in order to give room to other tasks with higher priorities. In addition, our approach is based on the idea of feedback, which offers the possibility of taking at run time the

Figure 3.3 **Example of a scientific document in image format. (Extracted from Marmot dataset)**

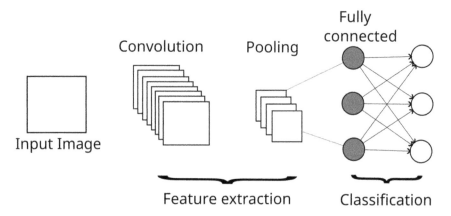

Figure 3.4 Basic architecture of a CNN with convolution layers, pooling layers, and fully connected layers. (Based on Khan et al. [13])

- Pooling layer: A layer that is used to reduce the size of convolutional feature maps produced by convolutional layers.
- Fully connected layer: A layer that is commonly used in the latter part of a neural network to process the high-level information extracted by the convolutional and pooling layers for image classification.

The most relevant operation in CNN is the convolution. A convolutional layer employs filters that are convolved with an input image, resulting in the creation of a feature map. Each filter that is used in a convolution layer is an array containing discrete numbers. For instance, the 3 × 3 filter as shown in Figure 3.5 does not have a defined initial form; the weights or numbers placed in each position of the matrix are initialized randomly. The weights of each filter are adjusted during a training process.

Filters used in the convolutional layer have adjustable hyperparameters, such as size (F) and stride (S), which indicates the number of pixels that the filter moves in each operation. There is a decrease in output image size following a convolution as demonstrated in Figure 3.6, necessitating the use of padding to increase the input size and ensure that the output image has the same dimensions as needed. Padding involves adding rows and columns with zeros around the input image. Considering that each convolutional layer consists of K filters, the feature map O is defined as follows:

$$O_k = f\left(b_k + \sum_c w_k[c] * I[c] \right)$$

(3.1)

-1	0	-1
0	5	0
3	-2	3

Figure 3.5 Convolutional filter of size 3 × 3, with random initialization.

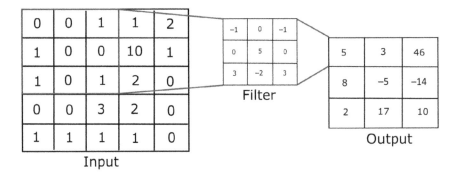

Input

Filter

Output

Figure 3.6 Example of a convolution operation with a 3 × 3 filter, stride of 1 pixel, and zero padding.

where f is the activation function, b_k is the k-bias, $I[c]$ is the input image on the c-channel, $w_k[c]$ is the k-filter for each channel, and $*$ represents convolution.

To carry out the convolution, the filter moves across the input image and computes the corresponding value in the output feature map. Beginning from the (0,0) position of the input image, the 3 × 3 filter is convolved within an area of equal size, and the resulting values are combined to produce the feature map data. The output feature map's dimensions are determined by the dimensions of the convolution filter, with height h and width w, and a stride size of S given by:

$$h' = \left\lceil \frac{h - f + s}{s} \right\rceil \tag{3.2}$$

$$w' = \left\lceil \frac{w - f + s}{w} \right\rceil \tag{3.3}$$

If a nonzero padding p is applied, then the equation is as follows:

$$h' = \left[\frac{h - f + s + p}{s} \right] \tag{3.4}$$

$$w' = \left[\frac{w - f + s + p}{w} \right] \tag{3.5}$$

By applying a zero padding, the feature map undergoes a reduction in dimension. To maintain the original input image size, it is advisable to use a padding of one.

The pooling layer is implemented on an input feature map typically next to a convolution layer decreasing the input feature map size. This operation is defined by a pooling function; the most used are average and max pooling as illustrated in Figure 3.7. The max-pooling function takes the maximum value of the input image within the window size, while the average-pooling function computes the average value of the window region over the input image. The result of this pooling layer is a downscaled feature map. Similar to the convolution layer, specifying the pooled region size and stride is crucial. The window scrolls through the input feature maps, with the size of the pooled region being fxf and a step of S. Equations 3.2 and 3.3 determine the output feature map size.

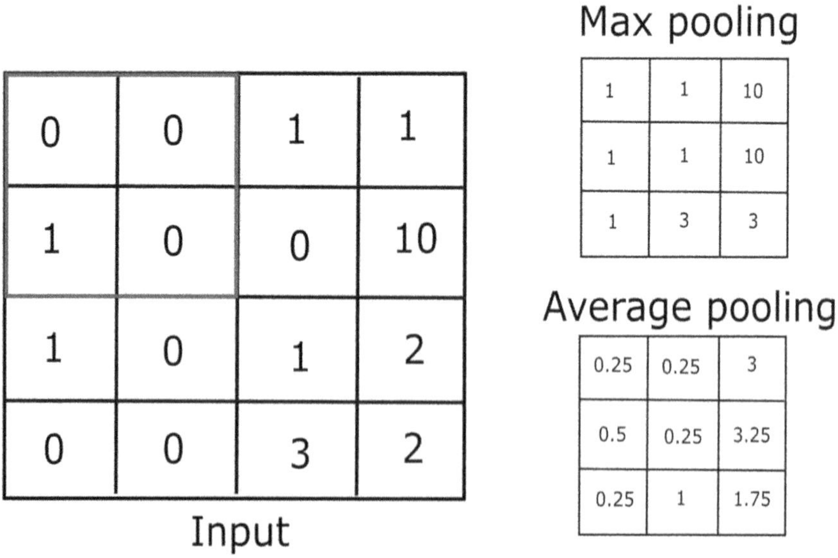

Max pooling

1	1	10
1	1	10
1	3	3

0	0	1	1
1	0	0	10
1	0	1	2
0	0	3	2

Input

Average pooling

0.25	0.25	3
0.5	0.25	3.25
0.25	1	1.75

Figure 3.7 Max-pooling and average-pooling operations of 2 × 2 dimension applied to a 4 × 4 input image with a stride of 1.

In neural networks, in general, the activation function is equivalent to the electrical impulse of a biological neuron. Typically, convolutional and fully connected layers are accompanied by nonlinear activation functions. The activation functions that are widely used in deep neural networks are the following:

a. The Sigmoid function: This function returns a value close to 0 for small values and close to 1 for large values.

$$f\,sigm(x) = \frac{1}{1+e^{-x}} \tag{3.6}$$

b. The Tanh function: This function applies the hyperbolic tangent function to the input values, scaling them to a range of [−1, 1].

$$f\,tan(x) = \frac{e^x - e^{-x}}{e^x + e^{-x}} \tag{3.7}$$

c. The Rectifier Linear Unit (ReLU) function: This function assigns a value of 0 to the input if it is negative while preserving its value if it is positive.

$$f\,ReLU(x) = \max(0,x) \tag{3.8}$$

d. The Leaky ReLU function: In this function, if the input is negative, the output is not completely turned off as in the traditional ReLU, but rather, a reduced version of the input is generated by multiplying the negative values by a rectifying coefficient.

$$fl-ReLU(x) = f(x) = \begin{cases} x\,if\,x > 0 \\ cx\,if\,x < 0 \end{cases} \tag{3.9}$$

e. The SoftMax function: This function transforms a vector with real values into a probability vector. This is commonly applied in the output layer of a neural network to obtain the classification, and since it is based on probabilities, the sum of the values in the output vector must be equal to 1.

$$f\,Softmax(x) = \frac{e^x}{\sum_{i=1}^{n} e^{x_i}} \tag{3.10}$$

Fully connected layers in CNNs are typically positioned at the end of the architecture. Every neuron in a fully connected layer is linked to all neurons in the preceding layer. The outcome of the convolutional or pooling layer is compressed into a single vector, where each value signifies the likelihood that a feature belongs to a specific label. The fully connected layer performs a simple matrix multiplication, followed by bias term addition and the application of a nonlinear function. The weight matrix for connections between neurons is denoted as W; the input and output activation vectors are represented as x and y, respectively; and b represents the vector of bias terms.

$$y = f(w^t x + b) \tag{3.11}$$

3.2.3 Quantum Convolutional Neural Networks

Classical computers use bits to store and process information, where each bit can have a state of either 1 or 0. In contrast, quantum computing uses qubits, which represent the state of a quantum particle. Due to the phenomenon of superposition, a qubit can be in a state of 1, 0, or any number in between. For instance, in a classical computer, two bits can store four possible values, such as 00, 01, 10, and 11, but only one at a time. However, in quantum computing, two qubits in superposition can represent the same four values simultaneously, as each qubit can be 1, 0, or both [26]. With each additional qubit, the number of possible values that can be represented grows exponentially.

The states $|0\rangle$ and $|1\rangle$ (Dirac notation) are known as computational basis states, defined as:

$$|0\rangle = \begin{bmatrix} 1 \\ 0 \end{bmatrix} \qquad\qquad (3.12)$$

$$|1\rangle = \begin{bmatrix} 0 \\ 1 \end{bmatrix} \qquad\qquad (3.13)$$

The computational basis for states involving two qubits is created by combining the tensor products of states involving a single qubit as follows:

$$|0\rangle \otimes |0\rangle = |0\rangle\,|0\rangle = |00\rangle = \begin{bmatrix} 1 \\ 0 \end{bmatrix} \otimes \begin{bmatrix} 1 \\ 0 \end{bmatrix} = \begin{bmatrix} 1 \\ 0 \\ 0 \\ 0 \end{bmatrix} \qquad\qquad (3.14)$$

Quantum states for multiple qubits are constructed by taking tensor products of individual qubit states. To illustrate, suppose we want to encode the bit string 11 on a quantum computer. We can do this by:

$$3 = 11 = \begin{bmatrix} 0 \\ 1 \end{bmatrix} \otimes \begin{bmatrix} 0 \\ 1 \end{bmatrix} = \begin{bmatrix} 0 \\ 0 \\ 0 \\ 1 \end{bmatrix} \qquad\qquad (3.15)$$

The probability of a qubit collapsing into either 1 or 0 is determined by its specific configuration. A qubit can be any linear combination (superpositions) of the states $|0\rangle$ and $|1\rangle$, and the state of a qubit can be represented using the following mathematical expression:

$$|\Psi\rangle = \alpha|0\rangle + \beta|1\rangle \qquad\qquad (3.16)$$

where α and β are two complex numbers that satisfy $|\alpha|^2 + |\beta|^2 = 1$. A measurement in quantum computing refers to the process of "observing" or "examining" a qubit, resulting in the collapse of its quantum state into one of the two possible classical states $\begin{bmatrix} 1 \\ 0 \end{bmatrix}$ or $\begin{bmatrix} 0 \\ 1 \end{bmatrix}$. When a given qubit is measured by the quantum state vector $\begin{bmatrix} \alpha \\ \beta \end{bmatrix}$, the result obtained is 0 with probability $|\alpha|^2$ and the outcome of 1 with probability $|\beta|^2$. The term used to describe this type of measurement is called z-measurement, although x- and y-measurements are more commonly used. The x-measurement and y-measurement are defined as follows, respectively:

$$|\pm\rangle := \frac{1}{2}(|0\rangle \pm |1\rangle) \tag{3.17}$$

$$|\pm i\rangle := \frac{1}{2}(|0\rangle \pm i|1\rangle) \tag{3.18}$$

It has been proven that any pure state $|\Psi\rangle$ that is normalized can be expressed in a three-dimensional space as a unit vector, such as the Bloch sphere:

$$|\psi\rangle := \cos\left(\frac{\theta}{2}\right)|0\rangle + e^{i\phi}\sin\left(\frac{\theta}{2}\right)|1\rangle \tag{3.19}$$

where $\theta \in [0, \pi]$ determines the probability of measuring $|0\rangle$ or $|1\rangle$ and $\varphi \in [0, 2\pi]$ describes the relative phase. As a result, it is possible to represent any state as a unit vector in the Bloch sphere, which is illustrated in Figure 3.8.

Like classical computers, quantum computers also consist of quantum circuits containing basic quantum gates to alter the state of a quantum system [28]. The commonly used symbols for quantum gates are shown in Figure 3.9. These quantum gates can be expressed using unitary matrices as follows:

$$|\psi'\rangle := U|\psi\rangle \tag{3.20}$$

This equation represents the operation of a quantum gate on a quantum state, where the gate is represented by a unitary matrix denoted as U.

Multiple single-qubit gates exist, and one of the most significant among them is the Hadamard gate. This gate is crucial as it helps to produce superpositions of $|0\rangle$ and $|1\rangle$, which then become the states $|+\rangle$ and $|-\rangle$, correspondingly.

$$H = \frac{1}{\sqrt{2}}\begin{bmatrix} 1 & 1 \\ 1 & -1 \end{bmatrix} \tag{3.21}$$

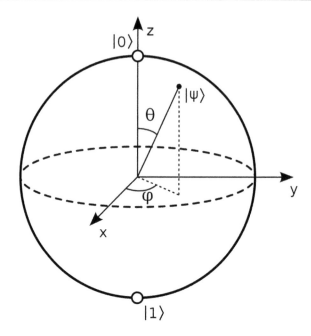

Figure 3.8 The Bloch sphere to visualize a single-qubit state. (Based on Wie [27])

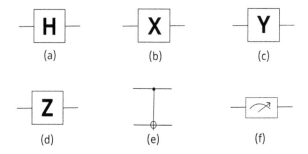

Figure 3.9 Notation of quantum gates. (a) Hadamard gate, (b) X-Pauli gate, (c) Y-Pauli gate, (d) Z-Pauli gate, (e) CNOT gate, and (e) Measurement gate.

The Pauli gates, X, Z, and Y, are single-qubit gates that operate on a qubit and are used to "negate," "phase change," and "negate with phase change" the state vector around the respective axis by π radians.

$$X = \frac{1}{\sqrt{2}} \begin{bmatrix} 0 & 1 \\ 1 & 0 \end{bmatrix} \tag{3.22}$$

$$Y = \frac{1}{\sqrt{2}} \begin{bmatrix} 0 & -i \\ -i & 0 \end{bmatrix} \tag{3.23}$$

$$Z = \frac{1}{\sqrt{2}}\begin{bmatrix} 1 & 0 \\ 0 & -1 \end{bmatrix} \qquad (3.24)$$

The CNOT gate is a two-qubit gate that works on a control qubit and a target qubit and acts like a switch. If the control qubit is set to 1, the value of the target qubit is flipped. This can be expressed as:

$$CNOT = \begin{bmatrix} 1 & 0 & 0 & 0 \\ 0 & 1 & 0 & 0 \\ 0 & 0 & 0 & 1 \\ 0 & 0 & 1 & 0 \end{bmatrix} \qquad (3.25)$$

A quantum circuit is composed of a series of moments, and each moment contains a set of operations that act on specific qubits. These operations are typically in the form of gates. In the circuit diagram, a solid line represents a qubit, with the top line being labeled as qubit register 0 and the rest being labeled sequentially [29]. Gates that act on one or more qubit registers are represented by boxes. The time flow in a quantum circuit is from left to right, and the gates are arranged chronologically, with the leftmost gate being applied to qubits first. Figure 3.10 shows the basic elements of a quantum circuit.

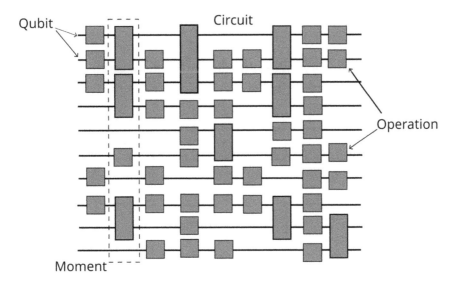

Figure 3.10 Classical quantum circuit illustrating the qubit, moment, and operation elements.

Variational Quantum Circuits (VQCs) are quantum algorithms that depend on free parameters [30–32]. It consists of three stages:

1. Preparation of an initial state: This step consists of setting all states of the qubit to zero.
2. A quantum circuit U(θ) parameterized by a set of free parameters θ: The first gates of the circuit can incorporate the state parameters (x) in a feature map, and the following gates use variational parameters (θ). Variation parameters are set to a random value.
3. Measurement: The measurement of the output at the end of the qubit wire is also called the expected value.

The variational parameters θ, along with an additional set of non-adaptive parameters $x = (x1, x2, \ldots)$, are processed as arguments to the gates of the circuit. This allows us to convert classical information (x, θ) into quantum information $U(x; \theta |0\rangle)$. In Figure 3.11, it is possible to observe a VQC of two qubits.

Ghosh et al. [25] introduced QCNNs. The QCNNs, which are based on CNNs, have a structure similar to that shown in Figure 3.12. Convolutions are operations that are performed on pairs of qubits, and they involve parameterized unitary rotations. After the convolution layers, there are pooling layers, which are performed by measuring a subset of the qubits and using the results to guide subsequent operations. Similar to classical CNNs, QCNNs are also applicable for image classification. In QCNNs, a multiqubit operation on the remaining qubits before the final measurement serves as the counterpart to a fully connected layer. Mathematically, a Quantum Layer (QL) can be defined as:

$$L(|\psi\rangle, \theta) : |\psi\rangle \rightarrow |\psi'\rangle = C(\theta)|\psi\rangle \tag{3.26}$$

where $C(\theta)$ is a VQC, θ represents the classical vector of weights to be optimized, and $|\Psi'\rangle$ is the output state.

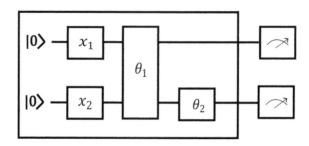

Figure 3.11 Variational quantum circuit of two qubits. (Based on McClean et al. [32])

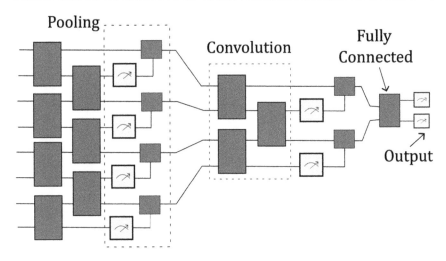

Figure 3.12 Structure of a QCNN. (Proposed by Ghosh et al. [25])

To enter the classical data in the quantum network (QN), it is necessary to encode it (φ) in the first quantum layer, and in the measurement, it is necessary to decode it (φ̂). A QN can be written as:

$$Q = \phi \circ Q \hat{\phi} \tag{3.27}$$

3.3 PROPOSED METHOD FOR CLASSIFYING TABLES

3.3.1 Hybrid Quantum Convolutional Neural Networks

In this study, a HQCNN is proposed for the classification of tables. This hybrid approach, introduced by Mari et al. [20] and modified by Ovalle-Magallanes et al. [23], consists of the following:

- A classical network (backbone network)
- A reduction layer due to the limitation of the number of qubits to process
- An L2-tanh layer to produce normalized output values
- A QN to generate HTL
- A Classic SoftMax layer for table or non-table classification

From the CNN, an output feature vector is produced, which is pre-processed before entering the QN, given that the QN's size is influenced by the quantity of characteristics that require processing. A D-VQC (variational distributed quantum circuit) is employed to execute the task using compact

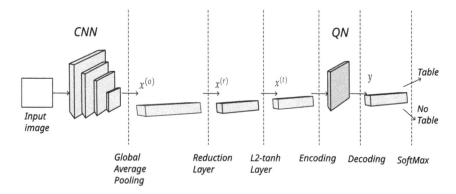

Figure 3.13 HQCNN architecture. (Proposed by Ovalle-Magallanes et al. [23])

quantum circuits. Figure 3.13 shows the architecture of the HQCNN with each of the stages. Moreover, Algorithm 1 in the Appendix includes a summary of the essential functions used in the HQCNN framework. The number of iterations is determined by the number of epochs required to train the complete architecture.

3.3.1.1 Classical Network

The Residual Network (ResNet) is a deep learning model presented by He et al. [33]. Residual blocks, which are an integral component of the ResNet model, have helped to deal with the difficulty of training extremely deep networks. Figure 3.14 illustrates that the model has a connection that skips certain layers, known as a skip connection, which is the fundamental concept behind residual blocks. The input X is multiplied with the layer weights and added with a bias term, but without the skip connection, this process would not occur.

The input $x(l)$ is added to the output $x(l + 1)$ of a short circuit layer block, as follows:

$$X^{l+1} = F\left(x^{(l)}, W^{(l)}\right) + x^{(l)} \tag{3.28}$$

where $F\left(x^{(l)}, W^{(l)}\right)$ refers to the residual mapping that requires training, and $W^{(l)}$ contains the weights of the l-layer. The ResNet-18 model comprises four residual blocks, using 64, 128, 256, and 512, respectively. Each convolutional layer within these blocks uses 3×3 filters and a stride size of two pixels.

3.3.1.2 Reduction Layer

Non-quantum data must be encoded in a QN in a space of dimension smaller than a CNN. However, real quantum computers provide a small number of

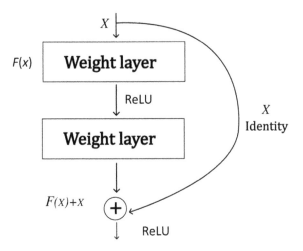

Figure 3.14 Residual block structure of a ResNet network.

qubits (e.g., two or four qubits); therefore, the use of raw images as an input for a quantum network is still infeasible, and they must be reduced in dimensions. Taking this into consideration, a CNN is used to obtain a feature vector $x^{(o)} \in R^{512}$, where $x^{(o)}$ refers to the output of the global average pooling layer in the ResNet-18 network. A linear reduction layer is used to reduce the feature vector to a $x(r) \in R^4$ vector, which serves as an input for the quantum network. The reduction layer is defined as:

$$L_r\left(x^{(o)}\right): x^{(o)} \rightarrow x^{(r)} = W^{(r)}x^{(o)} + b^{(r)} \tag{3.29}$$

where $W^{(r)}$ is a 4×512 matrix and $b^{(r)}$ is the bias vector.

3.3.1.3 L2-tanh Layer

Quantum coding methods consider that the input data is normalized and falls within a specific range. Nevertheless, the output generated by a CNN is not always normalized when connected to the inputs of a QN. Hence, another layer is added, the L2-tanh layer, which involves the inclusion of a scalar parameter k to prevent the oversaturation that can arise from a tanh function within the range of $[-1,1]$ (e.g., $\tanh(\pm 5) = \pm 0.999909 \approx \pm 1$). As a result, L2-tanhx(t) is calculated as:

$$x^{(t)} = f\left(k, x^{(r)}\right) = \tanh\left(k\frac{x^{(t)}}{\left\|x^{(r)}\right\|_2}\right) \tag{3.30}$$

where during the optimization process, k can be trained to regulate the saturation interval, with $x^{(r)}$ representing the reduced feature vector representation. The activation function no longer possesses a saturable section with the implementation of the L2-tanh layer.

3.3.1.4 Quantum Network

To distribute classical data across multiple VQCs in a QN, a distributed variational quantum circuit (D-VQC) is employed, where each VQC is independent. The decoded outputs of each VQC can then be concatenated. Assuming that n_x represents the size of classical data x, and n_q represents the number of VCQs with q qubits, it must satisfy the condition $n_x = n_q x q$. As a result, a quantum shell with D-VQC can be defined as:

$$L(|\Psi\rangle,\theta):|\Psi\rangle \rightarrow |\Psi'\rangle = \uplus C_i(\theta_i)|\Psi_i\rangle \tag{3.31}$$

where the operator $\uplus$ denotes concatenation, which is utilized to combine the output state $|\Psi'\rangle$ of each VQC into a single vector.

In this chapter, two- and four-qubit VQCs were implemented. A VQC of qi–qubits is defined as:

$$C = K \otimes_{k=1}^{q_i} R_i(\theta_k) \tag{3.32}$$

where K is obtained from the CNOT operation for each pair of consecutive qubits. K is defined as:

$$K = I\{k,k+1\} \otimes CNOT, \forall k \in q_i \tag{3.33}$$

The identity matrix is represented by I, and the indices k and $k+1$ denote the adjacent rows of the matrix, indicating the qubits where the CNOT gate is to be performed. In each VQC, the quantum embedding prepares each qubit in the unbiased state $|+\rangle$, which corresponds to a superposition of $|0\rangle$ and $|1\rangle$. The coding for the QN uses a single-qubit Hadamard gate, followed by a rotation of the Bloch sphere around the y-axis.

$$\phi(x)_k^{(t)} = \otimes_{k=1}^{q}\left(R_y\left(\frac{1}{2}\pi_k^{(t)}\right)H \right)|0_q\rangle \tag{3.34}$$

where $(x)^{(t)}$ is the output of the L2-tanh layer and $|0_q\rangle$ is the q-length initial state in a bit-string format (e.g., if $q = 2, |0_2\rangle = |00\rangle$).

After the quantum computation, the quantum measurement is obtained by evaluating the Pauli-Z observable, which generates an output between -1 and 1. This output is then mapped back to the classical data, and the resulting value is defined as the final output of the QN.

$$y = \hat{\Phi}(\Psi') = \Psi'|Z|\Psi'\rangle \tag{3.35}$$

3.3.2 Convolutional Neural Networks

As a point of comparison, five classical CNN architectures are trained. These CNNs are selected from the literature for their use in similar problems in UIE classification [10, 15, 17]. The architectures used are VGG-16, VGG-19, MobileNetV2, ResNet-50, and ResNet-18. The experiments were carried out with different training sets modified by data augmentation, in order to verify if the amount of data really influenced the result.

In the VGG networks, the idea of building a deep model was proposed by reusing simple basic blocks since they are of general use [34]. The VGG block comprises of a sequence of identical convolutional layers with a filter of size 3 × 3 and padding of 1, followed by a maximum pooling layer with a filter of size 2 × 2 and a stride of 2. The convolutional layer retains the input dimensions, while the pooling layer reduces them by half. The VGG network includes a module with a convolutional layer and a module with a fully connected layer. Several VGG blocks can be connected in a series by this module, with the number of layers being determined by hyperparameters. The variation between VGG-16 and VGG-19 is the number of layers in the network, with VGG-16 having 13 convolutional blocks and 3 fully connected layers, resulting in a depth of 16, while VGG-19 has 16 convolutional blocks and 3 fully connected layers, making a depth of 19. The architecture of both networks is illustrated in Table 3.1.

Considering the explanation of the residual blocks of ResNet networks in the previous section, the name of these networks varies depending on

Table 3.1 VGG Network Configuration, conv(receptive field size)-(channels). (Data from Simonyan et al. [34])

16 weight layers	19 weight layers
input (224 × 224 RGB image)	
conv3-64	conv3-64
conv3-64	conv3-64
MaxPool	
conv3-128	conv3-128
conv3-128	conv3-128
MaxPool	
conv3-256	conv3-256
conv3-256	conv3-256
conv3-256	conv3-256
	conv3-256
MaxPool	
conv3-512	conv3-512
conv3-512	conv3-512
conv3-512	conv3-512
	conv3-512

(Continued)

Table 3.1 (Continued)
VGG Network Configuration, conv(receptive
field size)-(channels). (Data from Simonyan et al.

16 weight layers	19 weight layers
MaxPool	
conv3-512	conv3-512
conv3-512	conv3-512
conv3-512	conv3-512
	conv3-512
MaxPool	
FC-(4096,4096,1000) × 3	
soft-max	

Table 3.2 ResNet Network Configuration (Data from He et al. [33])

18 weight layers	50 weight layers
7 × 7, 64, stride 2	
3 × 3 max pool, stride 2	
$\begin{bmatrix} 3x3, & 64 \\ 3x3, & 64 \end{bmatrix}$ x3	$\begin{bmatrix} 1x1, & 64 \\ 3x3, & 64 \\ 1x1, & 256 \end{bmatrix}$ x3
$\begin{bmatrix} 3x3, & 128 \\ 3x3, & 128 \end{bmatrix}$ x3	$\begin{bmatrix} 1x1, & 128 \\ 3x3, & 128 \\ 1x1, & 512 \end{bmatrix}$ x3
$\begin{bmatrix} 3x3, & 256 \\ 3x3, & 256 \end{bmatrix}$ x3	$\begin{bmatrix} 1x1, & 256 \\ 3x3, & 256 \\ 1x1, & 1024 \end{bmatrix}$ x3
$\begin{bmatrix} 3x3, & 512 \\ 3x3, & 512 \end{bmatrix}$ x3	$\begin{bmatrix} 1x1, & 512 \\ 3x3, & 512 \\ 1x1, & 2048 \end{bmatrix}$ x3
average pool, 1000-d fc, SoftMax	

the depth or number of layers they have. As shown in Table 3.2, from ResNet-50, each residual block is optimized, and the two 3 × 3 convolutional layers applied in ResNet-18 are replaced by 1 × 1 + 3 × 3 + 1 × 1. The middle 3 × 3 convolutional layer in ResNet-50 first reduces the computation under a reduced 1 × 1 convolutional layer and then restores it under another 1 × 1 convolutional layer, not only maintaining precision but also reducing the amount of computation [33].

MobileNetV1 is a general-purpose CNN designed for embedded devices, supporting classification and detection [35]. MobileNetV2 has made improvements on the basis of MobileNetV1, including classification, object detection, and semantic segmentation. MobileNetV2 builds on the concept of MobileNetV1 using a separable deep convolution as a block. In addition, V2 introduces two new features in the architecture:

- To enhance the range of features, a layer called "expansion" with a size of 1 × 1 was included prior to the deep convolution to boost the number of channels.
- On completion, the Linear function is used to prevent ReLU from destroying functions.

The Bottleneck residual blocks serve as the foundation of the architecture. These blocks take a tensor with k channels and perform three convolutions. The first convolution is a 1 × 1 pointwise convolution that expands the input feature map to a higher-dimensional space for nonlinear activations. The expansion factor is denoted as t, which results in t*k channels in the first step. Then, the ReLU6 activation function is applied. The second convolution is a depth convolution that utilizes 3 × 3 filters and is followed by ReLU6 activation. Finally, the filtered feature map is reprojected using another 1 × 1 pointwise convolution. However, the last step may result in the loss of information, which is why a linear activation function is applied. When the start and end feature maps have the same dimensions, a residual connection is added to facilitate gradient flow. The general structure of the MobileNetV2 network is shown in Table 3.4, where each line represents a sequence of one or more identical layers that are repeated n times. All layers

Table 3.3 Bottleneck Residual Block Structure. (Data from Sandler et al. [35])

Input	Operator	Output
h * w * k	1 × 1 conv2d, ReLU6	h * w * (tk)
h * w * tk	3 × 3 dwise s = s, ReLU6	h/s * w/s * tk
h/s * w/s * tk	linear 1 × 1 conv2d	h/s * w/s * k'

Table 3.4 MobileNetV2 Architecture Design. (Data from Sandler et al. [35])

Input	Operator	t	c	n	s
$224^2 * 3$	conv2d	-	32	1	2
$112^2 * 32$	bottleneck	1	16	1	1
$112^2 * 16$	bottleneck	6	24	2	2
$56^2 * 24$	bottleneck	6	32	3	2
$28^2 * 32$	bottleneck	6	64	4	2
$14^2 * 64$	bottleneck	6	96	3	1
$14^2 * 96$	bottleneck	6	160	3	2
$7^2 * 160$	bottleneck	6	320	1	1
$7^2 * 320$	bottleneck	–	1280	1	1
1*1*1280	conv2d 1 × 1	–	–	1	–
$224^2 * 1280$	avgpool 7 × 7	–	k	–	

in the same sequence produce the same number of output channels (c), and the first layer in each sequence has a stride of s, while the rest have a stride of 1. All convolutions in this network use 3 × 3 filters. Table 3.3 displays the structure of a residual bottleneck block.

3.3.3 Transfer Learning

As stated by Weiss et al. [14], transfer learning (TL) is employed to enhance learning in a target domain by transferring knowledge from a related domain. In simpler terms, this approach enables the transfer of problem-solving knowledge to tackle other problems. TL is used in deep learning because it requires large computation times and many resources. However, by using pre-trained models, TL allows complex problems to be solved in less time, taking advantage of the training for other case studies. Graphically, Figure 3.15 shows the transfer learning approach.

Pretrained models can be used as feature extractors where the architecture comprises layers of neurons that acquire different characteristics based on their depth. A fully connected layer is used to produce the final result, and its weights are frozen. The deeper the layer, the more it extracts specific features. For instance, to create a model that can recognize the plant species from an image, one can use the initial layers of a CNN model like AlexNet that was previously trained on ImageNet to classify 1,000 classes of images. Some of these classes may have characteristics that are helpful for the present objective, leveraging the prior learning.

Another TL technique is the adjustment of pretrained models. This is a more complex technique, in which not only the last layer is frozen for classification, but other layers are selectively retrained, as well. CNNs are highly configurable architectures with different hyperparameters. Moreover, while the first few layers capture general features, the last few layers focus primarily on the specific task at hand. Therefore, if the original purpose of training a network is not similar to the objective, it is more convenient to train more layers.

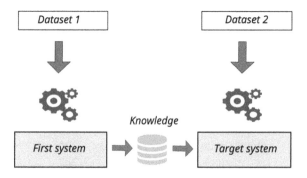

Figure 3.15 Transfer learning approach. (Based on Weiss et al. [14])

The process of classical to quantum transfer learning involves using non-quantum pretrained models to extract features, which are then processed in a quantum computer or a simulator. This approach is particularly useful for high-resolution imaging as quantum computers can handle only a limited number of features, making it more feasible than directly providing millions of raw pixels as an input to a quantum system [20].

This chapter uses pretrained models as feature extractors, where the weights of the pre-established output layer are maintained by freezing it. In addition, trainings are carried out from the scratch to compare the results of the TL. Moreover, through a HQCNN, the HTL is obtained showing the properties that quantum computing provides to the performance of the network.

3.3.4 Data Augmentation

When the dataset is inadequate or insufficient for a neural network to produce optimal results, data augmentation is applied. This involves using techniques to artificially enhance the dataset by creating new data from the original data through modifications [36]. Data can be modified or transformed, and deep learning models can also be used to generate new data. Data augmentation can be accomplished by applying a number of transformations, such as horizontal flipping, rotations, and random cropping to images, thus creating a new image that provides different information for the model to work with. When performing data augmentation, it is necessary to consider the specific application being used.

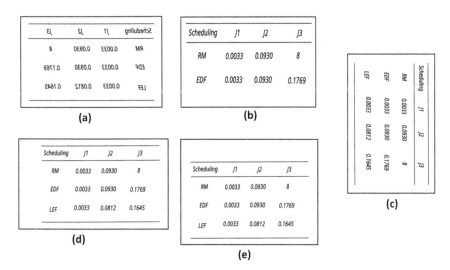

Figure 3.16 Implemented data augmentation: (a) Horizontal flipping, (b) Zoom in, (c) Rotation 90°, (d) Shifting image to the right, (e) Shifting image to the down.

The transformations used in this chapter are described next:

- Horizontal flipping (Figure 3.16 (a)): It refers to the axial symmetry of all the pixels taking the vertical line as the axis of symmetry that crosses the pixel.
- Zoom in (Figure 3.16 (b)): It involves resizing the image to a larger size than the original. In this case, each image was rescaled by 20%.
- Rotation 90° (Figure 3.16 (c)): This transformation is based on the angular displacement of all the pixels taking ($w/2$, $h/2$) as the central point, where w is the width and h is the height of the image.
- Shifting images right and down (Figure 3.16 (d, e)): The transformation makes the image move to the left, right, up, or down. In this chapter, the dataset is moved to the right and down. By shifting the original image in a particular direction, it is possible to fill the remaining space with a constant value; for instance, 0 or 255. This technique, referred to as padding, maintains the dimensions of the image following the transformation.

3.3.4.1 Implementation Details

Marmot dataset was used to train the classic networks, since it has images of tables and equations, although in this work, only the dataset of tables was used. Marmot dataset for tables has images as shown in Figure 3.3, images of complete scientific texts from which only the tables were cut for the classification test. Therefore, from Marmot, a dataset was generated whose images only contain the table as shown in Figure 3.1.

The new dataset included 2,200 training images, 276 for validation and 276 for testing. To train the networks, it was considered to apply augmentation (transformations to the images) of data to the images so that the dataset was larger and overfitting could be avoided. The data augmentation was carried out with the help of the Keras tool for Python, applying the following transformations as shown in Table 3.5: horizontal flipping, zoom in 20%, rotation 90 grades right, shifting right and shifting down.

Table 3.5 Data Augmentation Applied in Marmot Dataset.

ID	Accumulated number of images	Transformation
T1	2,200 Train, 276 Val/Test	Original
T2	4,400 Train, 552 Val/Test	Horizontal flipping
T3	4,400 Train, 552 Val/Test	Zoom in 20%
T4	4,400 Train, 552 Val/Test	Rotation 90 grades right
T5	4,400 Train, 552 Val/Tes	Shifting right
T6	4,400 Train, 552 Val/Test	Shifting down

3.3.5 Hyperparameters

The hyperparameters of a CNN are the values of the configurations used to train the network. The optimal value for each hyperparameter for a particular case study is difficult to know. Initially, it is necessary to use generic values or those values that have worked before in similar problems, or it is even possible to find the best values by testing the CNN. The hyperparameters that are considered for this work are the following:

- Pretrained: It determines if the network starts in a pretrained way or not and, therefore, if there is TL or not.
- Learning rate: This is a hyperparameter that plays a crucial role in an optimization algorithm by controlling the step size in each iteration toward the minimum of a loss function.
- Momentum: In gradient descent, this hyperparameter speeds up the search progress to traverse flat areas and smooth gradients.
- Batch size: It is the number of training samples before the model is updated.
- Epoch: It is the number of complete iterations during the training dataset.
- Optimizer: It indicates the type of optimizer for the gradient.
- Factor: It defines the factor by which the learning rate is reduced.
- Patience: This is the number of epochs without improvement after which the learning rate is reduced.
- Quantum network: It indicates if the quantum network is used or not.
- L2 normalization: It indicates whether or not an L2 normalization layer is used.
- Number of qubits: It represents the number of qubits to use in case of implementing the quantum network.
- Number of VQCs: It represents the number of quantum variational circuits used by the quantum network.
- Quantum depth: It defines the VQC Depth.

3.3.6 Evaluation Metrics

A binary classification task is carried out in this work where the result of the classification task can be either a "table" or a "non-table," considering some evaluation metrics that are used for this case study. These metrics include Accuracy, Precision, Recall, and F-Measure. Metrics are based on four representative detection results:

- True positive (TP): The test predicted a positive result, and the true value is also positive.
- True negative (TN): The test predicted a negative result, and the true value is also negative.

- False Negative (FN): The test predicted a negative result, but the true value is positive.
- False positive (FP): The test predicted a positive result, but the true value is negative.

Accuracy refers to the proportion of correct predictions over the number of observations.

$$Accuracy = \frac{TP + TN}{TP + TN + FP + FN} \times 100\% \tag{3.36}$$

Precision is calculated by dividing the number of correct predictions by the total number of predictions. This metric evaluates the accuracy of the correct data and compares it to the overall data returned.

$$Precision = \frac{TP}{TP + FP} \times 100\% \tag{3.37}$$

Recall metric reports how much the machine learning model can identify.

$$Recall = \frac{TP}{TP + FN} \times 100\% \tag{3.38}$$

F-Measure is calculated by taking the harmonic mean between precision and recall.

$$F_{\beta} = \left(\left(1 + \beta^2 \right) \frac{Precision * Recall}{\beta^2 \left(Precision * Recall \right)} \right) 100\% \tag{3.39}$$

The value of β is used to adjust the importance given to precision and recall in the F-Measure calculation. When β equals 1, the F-Measure is specifically referred to as F1-score.

3.4 COMPUTATIONAL EXPERIMENTS

To carry out the CNN and HQCNN experiments, a computer with the following characteristics was used: 128 GB RAM memory, Intel Xeon Silver 4214 processor, and NVIDIA Titan RTX 24 GB graphics card. The code was implemented in the programming language Python due to its extensive documentation and libraries available that allow using different open-source resources applicable to artificial intelligence. The libraries used in this case study are as follows:

a. For visualization purposes:
  - Matplotlib: It allows to create both digital and printed graphics.
  - Seaborn: This library is oriented to the visualization of statistical data.
b. For numerical calculation and data analysis:
  - NumPy: This library enables the use of high-dimensional vectors and matrices with vast amounts of data, which is useful for analyzing and interacting with algorithms.
c. For machine learning models:
  - Scikit-learn: It is used for supervised machine learning and model evaluation providing the evaluation metrics.
d. For using deep learning models:
  - TensorFlow: This is a tool used to construct and train neural networks. It works by encoding a graph as a data flow diagram for numerical calculations. The mathematical operations performed are represented by nodes in the graph, while tensors, which are arrays of multidimensional data, are represented by edges.
  - Keras: It is a high-level interface suitable for quickly testing whether the neural network developments will yield the intended results.
  - PyTorch: It is an open-source deep learning library built on top of the Torch library and used for various applications that involve deep learning.
e. For using quantum computing:
  - PennyLine: It is a Python library that enables programming of quantum computers across different platforms. It allows for the simulation of quantum computer training, similar to how a classical neural network is trained.

The model was validated using an 80–10–10 data split ratio for training, validation, and testing, respectively. As Table 3.5 shows, the dataset generated from the Marmot dataset contains 2,200 training images (1,100 tables and 1,100 non-tables) and 276 validation and testing images (138 tables and 138 non-tables), respectively. This dataset was modified with different transformations to check the performance with the original or increased number of images in CNNs. All images have a different original size as they are cutouts of a larger image, so each image was adjusted to a size of 224 × 224 pixels using python. An example of the transformation code can be found in Algorithm 2 in the Appendix. The hyperparameters for the experiments are initialized as shown in Table 3.6.

3.4.1 CNN Experiments

The tests with the different CNNs were carried out with the same training, validation, and testing data. In addition, tests in Table 3.5 were carried

Table 3.6 Initial Values of the Hyperparameters of the CNNs.

Hyperparameter	Value
Pretrained	True
Learning rate	0.001
Momentum	0.8
Batch size	4
Epochs	100
Optimizer	Adam
Patience	20
Factor	0.1
Only for QN	
Quantum	False
L2-tanh	N/A
Qubits	N/A
Number of VQCs	N/A
VQC Depth	N/A

out with the transformations separately. All the network architectures were trained using the same hyperparameters. The image pixel values were normalized on a linear scale to reduce the range from $[0, 255]$ to $[0, 1]$. During the training process, the adjustment parameters included a batch size of 4, and 100 epochs were used. The CNNs are used as a feature extractor, freezing all layers except the last one to use the weights that have been obtained from ImageNet pretraining. Algorithm 3 in the Appendix can be consulted to see how the process was carried out. The results of the network training with the different tasks are shown in Tables 3.7–3.12.

In general, as it can be seen, in the tables of results of the CNNs where the transfer learning is used, it is not enough to have an optimal performance. This may be because CNNs were not trained for a similar purpose. However, with the results it can be analyzed that CNN could be used as a starting point. In Table 3.7, with the original dataset, VGG-16 has the highest F1-score with 52.53% followed by ResNet-18 with 52.27%. In Table 3.8, adding the horizontal flipping transformation, ResNet-18 had the highest F1-score with 54.73% and then VGG-16 with 52.7%. In Table 3.9, adding a Zoom in of 20% the best performance in F1-score was MobileNetV2 with 54.23% and the second best VGG-16 with 51.43%. In Table 3.10, the best performances were from MobileNetV2 and VGG-16 with 49.49% and 48.82%, respectively. In Table 3.11, ResNet-18 had the best performance with an F1-score of 53.69% followed by VGG-16 with 52.13%. Finally, in Table 3.12, VGG-19 has the best F1-score with 53.98% followed by ResNet-18 with 50.13%. Although not all the transformations were effective in all the tests, the networks that obtained the best performance were VGG-16 and ResNet-18; in fact, in the T2 task, ResNet-18 obtained the best performance of all the tests, but still the results are not good.

Table 3.7 Networks Training Results with Using the Original Dataset.

		T1 original dataset			
Network	Time	Accuracy	Precision	Recall	F1-score
MobileNetV2	53min 8s	0.5217	0.5218	0.5217	0.5215
VGG-16	93min 10s	0.5254	0.5254	0.5254	0.5253
VGG-19	94min 24s	0.4964	0.4964	0.4964	0.4964
ResNet-50	64min 12s	0.5	0.5	0.5	0.4993
ResNet-18	60min 55s	0.529	0.5295	0.529	0.527

Table 3.8 Network Training Results Using the Original and the Horizontal Flipping Transformation Datasets.

		T2 Horizontal flipping			
Network	Time	Accuracy	Precision	Recall	F1-score
MobileNetV2	55min 32s	0.4928	0.4927	0.4928	0.4923
VGG-16	96min 14s	0.529	0.5295	0.529	0.527
VGG-19	95min 19s	0.4746	0.4746	0.4746	0.4745
ResNet-50	66min 50s	0.4891	0.4891	0.4891	0.4891
ResNet-18	62min 50s	0.5507	0.5523	0.5507	0.5473

Table 3.9 Network Training Results Using the Original and the Zoom in Transformation Datasets.

		T3 Zoom in 20%			
Network	Time	Accuracy	Precision	Recall	F1-score
MobileNetV2	54min 58s	0.5435	0.5439	0.5435	0.5423
VGG-16	95min 39s	0.5145	0.5145	0.5145	0.5143
VGG-19	94min 13s	0.4746	0.4738	0.4746	0.4703
ResNet-50	67min 8s	0.4928	0.4927	0.4928	0.4926
ResNet-18	62min 34s	0.4601	0.4571	0.4601	0.4503

Table 3.10 Network Training Results Using the Original and the Rotation 90° Rotation to the Right Transformation Datasets.

		T4 rotation 90° right			
Network	Time	Accuracy	Precision	Recall	F1-score
MobileNetV2	53min 59s	0.4964	0.4963	0.4964	0.4949
VGG-16	95min 2s	0.4928	0.4925	0.4928	0.4882
VGG-19	97min 18s	0.4855	0.4853	0.4855	0.4833
ResNet-50	68min 3s	0.4855	0.4839	0.4855	0.4721
ResNet-18	61min 15s	0.4819	0.4806	0.4819	0.4734

Table 3.11 Network Training Results Using the Original and the Shifting Right Transformation Datasets.

		T5 shifting right			
Network	Time	Accuracy	Precision	Recall	F1-score
MobileNetV2	56min 2s	0.4819	0.4819	0.4819	0.4817
VGG-16	94min 55s	0.5217	0.5218	0.5217	0.5213
VGG-19	95min 45s	0.5109	0.5113	0.5109	0.5061
ResNet-50	65min 23s	0.4674	0.4673	0.4674	0.4672
ResNet-18	61min 12s	0.5471	0.5517	0.5471	0.5369

Table 3.12 Network Training Results Using the Original and the Shifting Left Transformation Datasets.

		T6 shifting left			
Network	Time	Accuracy	Precision	Recall	F1-score
MobileNetV2	56min 10s	0.4746	0.4746	0.4746	0.4746
VGG-16	94min 22s	0.4674	0.4674	0.4674	0.4673
VGG-19	95min 8s	0.5399	0.5399	0.5399	0.5398
ResNet-50	66min 6s	0.4819	0.4819	0.4819	0.4817
ResNet-18	62min 29s	0.5036	0.5037	0.5036	0.5013

Table 3.13 Results of the Networks Training with the Weights Initialized with Transfer Learning.

		T1 original dataset			
Network	Time	Accuracy	Precision	Recall	F1-score
MobileNetV2	26min 20s	0.906	0.8986	0.93	0.914
VGG-16	63min 10s	0.9275	0.9274	0.92	0.9237
VGG-19	64min 24s	0.9198	0.9274	0.9055	0.9163
ResNet-50	45min 7s	0.8818	0.8854	0.8782	0.8818
ResNet-18	26min 22s	0.9348	0.942	0.9285	0.9352

For subsequent experiments, the weights were initialized with pretraining the networks. During the training process, Stochastic Gradient Descent was used with a momentum of 0.8 (Stochastic Gradient Descent with Momentum [SGDM]), using a learning rate of 0.001, a patience of 20, and a batch size of 4. The training process continued for 100 epochs. ResNet-18 had the best F1-score performance in the CNN tests with the T2 task, so it is reused for HQCNN with the same dataset. Since there is not much difference between the original dataset and those with transformations, the original is used. The outcomes of the transfer learning training are illustrated in Table 3.13.

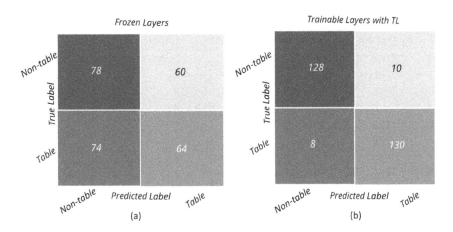

Figure 3.17 ResNet-18 confusion matrices: (a) frozen layers and only last trainable layer, (b) all trainable layers with TL to initialize the weights.

By unfreezing all the layers but keeping the initial weights from pretraining, a significant enhancement is observed in the performance of the networks. In these experiments, the CNNs no longer keep the weights trained with ImageNet but take them as a starting point to modify them for the classification of tables. The ResNet-18 network has a better performance together with VGG-16; the difference is the execution time since the VGG-16 networks are considerably more computationally demanding. The discrepancy in the confusion matrices of ResNet-18 with frozen layers versus trainable layers is depicted in Figure 3.17. The distribution of true positives and true negatives is more stable with trainable layers, since without them more false cases are detected.

3.4.2 HQCNN Experiments

The hyperparameters for all network configurations were the same. The values of every pixel were normalized from the original range from 0–255 to 0–1. The option to use pretraining for classic TL was both enabled and disabled. The optimization method used was SGDM, with a learning rate of 0.001 and a momentum of 0.8. To prevent overfitting, the network weights were only updated if the validation loss decreased. If the validation loss did not improve for 20 consecutive epochs, the learning rate would decrease by a scaling factor of 0.1. The training process used a batch size of 4 for 100 epochs. The L2-tanh layer activation was switched depending on the number of qubits (two or four qubits) and the number of VQCs and their depth (1 or 2). The ResNet-18 model was used as the backbone network for the HQCNN architecture due to its optimal performance and characteristics. Table 3.14 shows the hyperparameter combinations that were used in the tests.

Table 3.14 Combination of Modified Hyperparameters in Training.

ID	Pretrain	Quantum	L2 layer	Qubits	VQCs	Q-depth
1	True	False	N/A	N/A	N/A	N/A
2	True	True	False	4	1	1
3	True	True	True	4	1	1
4	True	True	False	4	1	2
5	True	True	True	4	1	2
6	True	True	False	2	1	1
7	True	True	True	2	1	1
8	True	True	False	2	1	2
9	True	True	True	2	1	2
10	True	True	False	2	2	1
11	True	True	True	2	2	1
12	True	True	False	2	2	2
13	True	True	True	2	2	2
14	False	False	N/A	N/A	N/A	N/A
15	False	True	False	4	1	1
16	False	True	True	4	1	1
17	False	True	False	4	1	2
18	False	True	True	4	1	2
19	False	True	False	2	1	1
20	False	True	True	2	1	1
21	False	True	False	2	1	2
22	False	True	True	2	1	2
23	False	True	False	2	2	1
24	False	True	True	2	2	1
25	False	True	False	2	2	2
26	False	True	True	2	2	2

The classification results presented in Table 3.15 show that the use of HQCNN significantly improved the performance of classical networks. The best accuracy, precision, recall, and F1-score were achieved by the HQCNN architecture that included a pretrained network, an active L2-tanh layer, and a 1-layer deep QN using a four-qubit VQC. The accuracy was 0.9818%, precision was 0.9784%, recall was 0.9855%, and F1-score was 0.9819%. However, the same configuration without pretraining obtained an accuracy of 0.9528%, a precision of 0.9629%, a recall of 0.942%, and an F1-score of 0.9523%. In addition, the HQCNN architecture with only two qubits performed well, including a pretrained network, an active L2-tanh layer, and a 2-layer deep QN using a VQC, achieving an accuracy of 0.9891%, a precision of 0.9927%, a recall of 0.9855%, and an F1-score of 0.989%. The same configuration without pretraining obtained an accuracy of 0.9637%, a precision of 0.9848%, a recall of 0.942%, and an F1-score of 0.9629%.

Table 3.15 HQCNN Training Results with the Original and Horizontal Flipping Datasets.

ID	Time	Accuracy	Precision	Recall	F1-score
1	26min 22s	0.9348	0.942	0.9285	0.9352
2	89min 33s	0.967391	0.957447	0.978261	0.967742
3	90min 39s	0.981884	0.978417	0.985507	0.981949
4	114min 5s	0.981884	0.985401	0.978261	0.981818
5	111min 26s	0.974638	0.978102	0.971014	0.974545
6	63min 4s	0.981884	0.985401	0.978261	0.981818
7	60min 10s	0.985507	0.978571	0.992754	0.978261
8	71min 1s	0.971014	0.964286	0.978261	0.971223
9	72min 29s	0.98913	0.992701	0.985507	0.989091
10	110min 30s	0.978261	0.971429	0.985507	0.978417
11	112min 17s	0.981884	0.992593	0.971014	0.981685
12	129min 4s	0.971014	0.985075	0.956522	0.970588
13	133min 24s	0.974638	0.964539	0.985507	0.97491
14	26min 26	0.8429	0.84209	0.849	0.8455
15	88min 41s	0.952899	0.969925	0.934783	0.95203
16	89min 20s	0.952899	0.962963	0.942029	0.952381
17	113min 31s	0.963768	0.963768	0.963768	0.963768
18	111min 8s	0.927536	0.960938	0.891304	0.924812
19	85min 19s	0.945652	0.962406	0.927536	0.944649
20	61min 1s	0.952899	0.969925	0.934783	0.95203
21	93min 58s	0.934783	0.934783	0.934783	0.934783
22	73min 3s	0.963768	0.984848	0.942029	0.962963
23	129min 6s	0.934783	0.934783	0.934783	0.934783
24	112min 37s	0.952899	0.962963	0.942029	0.952381
25	150min 28s	0.949275	0.962687	0.934783	0.948529
26	133min 32s	0.934783	0.96875	0.898551	0.932331

In general, pretrained HQCNN setups performed better than those trained from the scratch. The L2-tanh layer did help to increase the network performance a bit, and the execution time difference is minimal in most cases. Training times increased considerably with HQCNNs but performance for both the cases (QCNN and HQCNN) are equal. Figure 3.18 shows the confusion matrices for the training of ResNet-18 with TL and ResNet-18 with HTL. The performance improvement provided by HTL is confirmed with the confusion matrices.

Table 3.16 shows a comparison between the proposed hybrid method and the classical or non-quantum methods in the literature to detect tables. It is appreciable that the proposal of this work surpasses the results of other methods, demonstrating the power of using hybrid quantum networks.

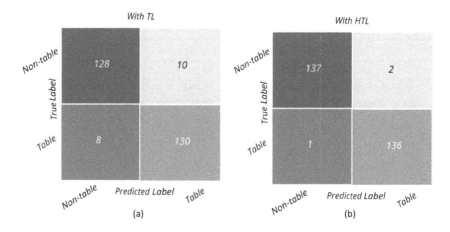

Figure 3.18 ResNet-18 confusion matrices: (a) results of applying TL, (b) results of applying HTL.

Table 3.16 Comparison of the Results in Marmot Dataset with the Literature.

Model	F1-score
DeCNT [8]	0.895
CDeC-Net [9]	0.952
HybridTabNet [10]	0.956
CasTabDetectoRS [11]	0.958
DCTable [12]	0.969
HQCNN	0.989

3.5 CONCLUSION AND FUTURE WORK

In this work, an HQCNN was implemented to demonstrate that the results are superior compared to a CNN. The proposed network structure was presented, and a number of combinations of hyperparameters were tested to see the changes in the results. Several experiments were performed with pretrained CNNs, and one of them (ResNet-18) was chosen for the non-quantum section of the structure.

The experiments were carried out under the same conditions; the performance of the CNNs was considerably better when the layers were enabled for their training. It is important to emphasize that the weights were not initialized randomly to take advantage of the TL of the ImageNet dataset with which the CNNs were trained. Being randomly initialized, it can be complex to achieve the performance obtained through TL, since ImageNet is a very large dataset, unlike its own. It is probably necessary for the future work to validate the dataset with more examples of tables and non-tables and thus ensure that the overfitting is minimal. The dataset was divided in a fixed way, so it would be interesting to try a cross-validation to see the difference in the results.

It was demonstrated that the HTL helped the table or the non-table classification performance. Using deep learning with quantum computing, the computational cost is higher than the tests with CNNs; however, the performance is higher. An inconvenience that may arise is that the tests were carried out in a quantum computer simulation using Python, where the conditions are stable. In a real quantum computer, this is not usually the case, because the hardware does not provide these types of conditions. For future investigations, it would be appropriate to test this proposal in a real environment where the quantum states are not ideal. There are some companies such as IBM that provide quantum hardware for testing, so it could be considered an option.

Quantum computing improves the overall table classification performance, so it can be considered to train an architecture that includes not only sorts of tables but also other UIEs. The limitation of QN is the number of qubits that can be used, at least with the language and the libraries used, so hybrid networks are very useful for these case studies with images of adequate dimensions for CNNs (e.g., 224 × 224 pixels). Therefore, considering this limitation, it was not possible to design and evaluate a QN without using hybrid networks.

For future developments, building a non-binary structure that has the ability to recognize tables and equations with optimal performance could be expected. Furthermore, it is expected to test this model with a new dataset, and the obtained results will be compared with those obtained by using the Marmot dataset.

Appendix

Algorithm 1 Hybrid Quantum Convolutional Neural Network

Require: CNN, N; QN, Q; dataset, D
Ensure: Trained HQCNN
1: **for** $e \leftarrow 0$ to *epochs* **do**
2:     **for** $I \in D$ **do**
3:        CNN feature extraction;
4:        $x^{(o)} \leftarrow N(I)$;
5:        Feature reduction;
6:        $x^{(r)} \leftarrow L_r(x^{(o)})$;
7:        Feature normalization;
8:        $x^{(t)} \leftarrow tanh\left(K \frac{x^{(r)}}{||x^{(r)}||_2}\right)$;
9:        QN processing;
10:       $y = Q\left(|\phi(x^{(t)})\rangle\right)$;
11:       SoftMax clasification layer;
12:       $\hat{p} = SoftMax(y)$;
13:     **end for**
14:     Backpropagation;
15: **end for**

Algorithm 2 Data generator and data augmentation

```
1  datagen = ImageDataGenerator(
2      rescale=1./255,
3      horizontal_flip=True,
4      zoom_range=0.2,
5      rotation_range=90,
6      width_shift_range=0.1,
7      height_shift_range=0.1,
8      preprocessing_function=preprocess_input)
9
10 train_generator = datagen.flow_from_directory(
11     train_data_dir,
12     target_size=(image_shape, image_shape),
13     batch_size=batch_size,
14     class_mode='categorical')
```

Algorithm 3 Data generator and data augmentation

```
1 #define pretrained model
2 model = applications.vgg16.VGG16(input_tensor=
3 image_input, include_top=True, weights='imagenet')
4 #Get last layer
5 last_layer = model.layers[-2].output
6 #modify last layer for two classes
7 out = Dense(num_classes, activation='softmax',
8 name='output')(last_layer)
9 custom_vgg_model = Model(image_input, out)
10 #Freeze all layers except the last one
11 for layer in custom_vgg_model.layers[:-1]:
12   layer.trainable = False
```

Algorithm 4 Network optimization with initialized weights using transfer learning

```
1 backbone = torchvision.models.resnet50(pretrained=True)
2
3 if PRETRAINED and not FINETUNE:
4     for param in backbone.parameters():
5         param.requires_grad = False
6
7 model = backbone.to(device)
8
9 criterion = nn.CrossEntropyLoss()
10 optimizer = optim.SGD(model.parameters(), lr=0.001,
11            momentum=0.8)
12 scheduler = lr_scheduler.ReduceLROnPlateau(optimizer=
13     optimizer, mode='min', factor=0.1,
14            patience=20, verbose=True)
```

Algorithm 5 General structure for HTL with ImageNet

```
1
2 # classical cnn parameters
3 # number of output features
4 self.n_qubits = n_qubits
5 self.num_residuals = num_residuals
6 self.num_filters = 512 #backbone.fc.in_features
7 # all conv layers, remove last (classification) layer
8 # split by conv1, resblock1,2,3, & gap
9 #layers = list(backbone.children())[:-1]
10 self.conv1 = nn.Sequential(*list(backbone.children())[0:4])
11 if num_residuals == 1:
12     self.resblock1 = nn.Sequential(*list(backbone.children())[4])
13     self.num_filters = 64
14 if num_residuals == 2:
15     self.resblock1 = nn.Sequential(*list(backbone.children())[4])
16     self.resblock2 = nn.Sequential(*list(backbone.children())[5])
17     self.num_filters = 128
18 if num_residuals == 3:
19     self.resblock1 = nn.Sequential(*list(backbone.children())[4])
20     self.resblock2 = nn.Sequential(*list(backbone.children())[5])
21     self.resblock3 = nn.Sequential(*list(backbone.children())[6])
22     self.num_filters = 256
23 if num_residuals == 4:
24     self.resblock1 = nn.Sequential(*list(backbone.children())[4])
25     self.resblock2 = nn.Sequential(*list(backbone.children())[5])
26     self.resblock3 = nn.Sequential(*list(backbone.children())[6])
27     self.resblock4 = nn.Sequential(*list(backbone.children())[7])
28     self.num_filters = 512
29 self.avgpool = nn.AdaptiveAvgPool2d((1, 1))
```

```
30
31 # number of classes
32 self.num_target_classes = num_target_classes
33 # pre-trained network
34 #self.backbone = backbone
35 # use l2-gap
36 self.use_l2 = use_l2
37 # use quantum dresses network
38 self.use_quantum = use_quantum
39 # feature extractor network
40 #self.feature_extractor = nn.Sequential(*layers)
41 # normalization layer in case of use_l2 = True
42 if use_l2:
43     self.normalization = NormLayer()
```

```
1 if use_quantum:
2 # quantum parameters
3     self.reduction = nn.Linear(self.num_filters, n_qubits*
    n_qlayers)
4     self.n_qlayers = n_qlayers
5     self.q_layers_list = list([QuantumLayer(q_depth=q_depth,
    n_qubits=n_qubits, q_delta=q_delta) for _ in range(n_qlayers)
    ])
6
```

```
7 # classification layer in case of use_quantum == True
8 self.q_classifier = nn.Linear(n_qubits * n_qlayers,
    num_target_classes)
9
10 else:
11 # classification layer in case of use_quantum == False
12     self.classifier = nn.Linear(self.num_filters,
    num_target_classes)
```

REFERENCES

[1] B. H. Phong, T. M. Hoang, and T. L. Le, "A Hybrid Method for Mathematical Expression Detection in Scientific Document Images," *IEEE Access*, vol. 8, pp. 83663–83684, 2020, doi: 10.1109/ACCESS.2020.2992067.

[2] J. Fang, X. Tao, Z. Tang, R. Qiu, and Y. Liu, "Dataset, Ground-truth and Performance Metrics for Table Detection Evaluation," in *Proceedings—10th IAPR International Workshop on Document Analysis Systems, DAS 2012*, 2012, pp. 445–449, doi: 10.1109/DAS.2012.29.

[3] W. Ohyama, M. Suzuki, and S. Uchida, "Detecting Mathematical Expressions in Scientific Document Images Using a U-Net Trained on a Diverse Dataset," *IEEE Access*, vol. PP, no. 8, p. 1, 2019, doi: 10.1109/ACCESS.2019.2945825.

[4] X. Lin, L. Gao, Z. Tang, and X. Lin, "Mathematical Formula Identification in PDF Documents," in *2011 International Conference on Document Analysis and Recognition*, 2011, pp. 1419–1423, doi: 10.1109/ICDAR.2011.285.

[5] P. Kukkadapu, M. Mahdavi, and R. Zanibbi, "ScanSSD : Scanning Single Shot Detector for Mathematical Formulas in PDF Document Images," in *Computer Vision and Pattern Recognition*, 2020, pp. 1–8.

[6] J. Luo, Z. Li, J. Wang, and C. Lin, "ChartOCR : Data Extraction from Charts Images via a Deep Hybrid Framework," in *ChartOCR: Data Extraction from Charts Images via a Deep Hybrid Framework*, 2021, pp. 1917–1925.

[7] Y. Liu, K. Bai, P. Mitra, and C. L. Giles, "TableSeer : Automatic Table Metadata Extraction and Searching in Digital Libraries," in *JCDL*, 2007, pp. 91–100.

[8] S. Siddiqui, M. Malik, S. Agne, A. Dengel, and S. Ahmed, "DeCNT : Deep Deformable CNN for Table Detection," *IEEE Access*, vol. 6, pp. 1–11, 2018, doi: 10.1109/ACCESS.2018.2880211.

[9] M. Agarwal and A. Mondal, "CDeC-Net : Composite Deformable Cascade Network for Table Detection in Document Images," in *Computer Vision and Pattern Recognition*, 2020, pp. 1–12.

[10] D. Nazir, K. A. Hashmi, A. Pagani, M. Liwicki, and M. Z. Afzal, "HybridTabNet: Towards Better Table Detection in Scanned Document Images," *Appl. Sci.*, vol. 11, pp. 1–22, 2021, doi: 10.3390/app11188396.

[11] K. A. Hashmi, A. Pagani, M. Liwicki, D. Stricker, and M. Z. Afzal, "CasTabDetectoRS : Cascade Network for Table Detection in Document Images with Recursive Feature Pyramid and Switchable Atrous Convolution," *J. Imaging*, vol. 7, no. 214, pp. 1–23, 2021, doi: 10.3390/jimaging7100214.

[12] T. Kazdar, W. S. Mseddi, M. A. Akhloufi, A. Agrebi, and M. Jmal, "DCTable : A Dilated CNN with Optimizing Anchors for Accurate Table Detection," *J. Imaging*, vol. 9, no. 62, pp. 1–22, 2023, doi: 10.3390/jimaging9030062.

[13] S. Khan, H. Rahmani, S. A. A. Shah, and M. Bennamoun, *A Guide to Convolutional Neural Networks for Computer Vision*, 1st ed., vol. 8. Morgan & Claypool, 2018.

[14] K. Weiss, T. M. Khoshgoftaar, and D. D. Wang, "A Survey of Transfer Learning," *J. Big Data*, vol. 3, no. 1, Dec. 2016, doi: 10.1186/s40537-016-0043-6.

[15] S. Schreiber, S. Agne, I. Wolf, A. Dengel, and S. Ahmed, "DeepDeSRT: Deep Learning for Detection and Structure Recognition of Tables in Document Images," *Proc. Int. Conf. Doc. Anal. Recognition, ICDAR*, vol. 1, pp. 1162–1167, 2017, doi: 10.1109/ICDAR.2017.192.

[16] M. Gobel, T. Hassan, E. Oro, and G. Orsi, "ICDAR 2013 Table Competition," in *12th International Conference on Document Analysis and Recognition ICDAR*, 2013, pp. 1449–1453, doi: 10.1109/ICDAR.2013.292.

[17] S. S. Paliwal, D. Vishwanath, R. Rahul, M. Sharma, and L. Vig, "TableNet: Deep Learning Model for End-to-end Table Detection and Tabular Data Extraction from Scanned Document Images," *Proc. Int. Conf. Doc. Anal. Recognition, ICDAR*, pp. 128–133, 2020, doi: 10.1109/ICDAR.2019.00029.

[18] J. Biamonte, P. Wittek, N. Pancotti, P. Rebentrost, N. Wiebe, and S. Lloyd, "Quantum Machine Learning," *Nature*, 2018, vol. 549, no. 7671, pp. 195–202, doi: 10.1038/nature23474.

[19] Y. Kwak, W. J. Yun, S. Jung, and J. Kim, "Quantum Neural Networks: Concepts, Applications, and Challenges," in *International Conference on Ubiquitous and Future Networks, ICUFN*, 2021, vol. 2021-August, pp. 413–416, doi: 10.1109/ICUFN49451.2021.9528698.

[20] A. Mari, T. R. Bromley, J. Izaac, M. Schuld, and N. Killoran, "Transfer Learning in Hybrid Classical-quantum Neural Networks," in *Quantum Physics*, 2020, vol. 4, doi: 10.22331/Q-2020-10-09-340.

[21] E. H. Houssein, Z. Abohashima, M. Elhoseny, and W. M. Mohamed, "Hybrid Quantum-classical Convolutional Neural Network Model for COVID-19 Prediction Using Chest X-ray Images," *J. Comput. Des. Eng.*, vol. 9, no. 2, pp. 343–363, 2022, doi: 10.1093/jcde/qwac003.

[22] V. Kulkarni, S. Pawale, and A. Kharat, "A Classical-Quantum Convolutional Neural Network for Detecting Pneumonia from Chest Radiographs," in *Computer Vision and Pattern Recognition*, 2022, pp. 1–15, [Online]. Available: http://arxiv.org/abs/2202.10452.

[23] E. Ovalle-Magallanes, J. G. Avina-Cervantes, I. Cruz-Aceves, and J. Ruiz-Pinales, "Hybrid Classical–quantum Convolutional Neural Network for Stenosis Detection in X-ray Coronary Angiography," *Expert Syst. Appl.*, vol. 189, no. September 2021, p. 116112, 2022, doi: 10.1016/j.eswa.2021.116112.

[24] A. Sebastianelli, D. A. Zaidenberg, D. Spiller, B. Le Saux, and S. Ullo, "On Circuit-Based Hybrid Quantum Neural Networks for Remote Sensing Imagery Classification," *IEEE J. Sel. Top. Appl. Earth Obs. Remote Sens.*, vol. 15, pp. 565–580, 2022, doi: 10.1109/JSTARS.2021.3134785.

[25] A. Ghosh, A. Sufian, F. Sultana, A. Chakrabarti, and D. De, *Fundamental Concepts of Convolutional Neural Network*, vol. 172, no. June, Springer, 2019.

[26] V. Kulkarni, M. Kulkarni, and A. Pant, *Quantum Computing Methods for Supervised Learning*, vol. 3, no. 2, Springer, 2021.

[27] C. R. Wie, "Two-Qubit Bloch Sphere," in *Quantum Physics*, 2020, vol. 2, no. 3, pp. 383–396, doi: 10.3390/physics2030021.

[28] P. Kaye, R. Laflamme, and Mi. MOsca, *An Introduction to Quantum Computing for Non-physicists*. Oxford University Press, 2007.

[29] R. de Wolf, "Quantum Computing: Lecture Notes," 2019, [Online]. Available: http://arxiv.org/abs/1907.09415.

[30] E. Farhi and H. Neven, "Classification with Quantum Neural Networks on Near Term Processors," pp. 1–21, 2018, [Online]. Available: http://arxiv.org/abs/1802.06002.

[31] S. Sim, P. D. Johnson, and A. Aspuru-Guzik, "Expressibility and Entangling Capability of Parameterized Quantum Circuits for Hybrid Quantum-Classical Algorithms," *Adv. Quantum Technol.*, vol. 2, no. 12, pp. 1–18, 2019, doi: 10.1002/qute.201900070.

[32] J. R. McClean, J. Romero, R. Babbush, and A. Aspuru-Guzik, "The theory of variational hybrid quantum-classical algorithms," *New J. Phys.*, vol. 18, no. 2, 2016, doi: 10.1088/1367-2630/18/2/023023.

[33] K. He, X. Zhang, S. Ren, and J. Sun, "Deep residual learning for image recognition," in *Proceedings of the IEEE Computer Society Conference on Computer Vision and Pattern Recognition*, 2016, pp. 1–12, doi: 10.1109/CVPR.2016.90.

[34] K. Simonyan and A. Zisserman, "Very Deep Convolutional Networks for Large-Scale Image Recognition," *Am. J. Heal. Pharm.*, vol. 75, no. 6, pp. 398–406, 2015.

[35] M. Sandler, A. Howard, M. Zhu, A. Zhmoginov, and L. C. Chen, "MobileNetV2: Inverted Residuals and Linear Bottlenecks," *Proc. IEEE Comput. Soc. Conf. Comput. Vis. Pattern Recognit.*, pp. 4510–4520, 2018, doi: 10.1109/CVPR.2018.00474.

[36] C. Shorten and T. M. Khoshgoftaar, "A Survey on Image Data Augmentation for Deep Learning," *J. Big Data*, vol. 6, no. 1, 2019, doi: 10.1186/s40537-019-0197-0.

Chapter 4

Transformation Optics
Subwavelength Control of Light Leads to Novel Phased Array Antenna System Design

Dipankar Mitra, Eric Jahns, Shuvashis Dey, and Sayan Roy

4.1 INTRODUCTION

Quantum computing has the potential to address many of the world's current challenges in such fields as agriculture, environment, health, energy, climate, and materials science, as well as others that are yet to emerge [1]. Classical computing often faces limitations when dealing with complex and large-scale systems. However, quantum systems, if designed to scale, are expected to surpass the capabilities of even the most powerful supercomputers. One important area of research in time-dependent quantum mechanics/optics is spectroscopy, which involves studying matter through its interaction with electromagnetic radiation [2]. In classical physics, light-matter interactions occur when an oscillating electromagnetic field resonantly interacts with charged particles. This interaction occurs over a time scale of $\tau \approx 10^{-15}$ s (where τ refers to the time constant) or a characteristic wavelength of $\lambda \approx 10^{-7}$ to 10^{-6} m [3]. Transformation optics (TO) [4, 5] has been a valuable tool since its introduction in 2006 for designing and understanding electromagnetic systems. It offers a geometric perspective of light waves without relying on a ray approximation.

TO presents a novel and systematic methodology for designing unconventional electromagnetic devices through the use of an appropriate coordinate transformation. The key assumption underlying this approach is the form-invariance of Maxwell's equations under coordinate transformations [6–8]. The desired material properties can be derived by starting with the familiar form of Maxwell's equations, and this methodology can be extended to arbitrary and complex geometries using the same basic principles [9]. The successful demonstration of two-dimensional invisibility cloaks in microwave frequencies [10, 11], utilizing state-of-the-art materials, has catalyzed research in designing novel devices using the TO technique. This field of research has progressed significantly as a result.

The TO approach has led to the design of numerous passive electromagnetic devices, such as the optical cloak with a reduced set of material parameters, all-dielectric cloak in the THz regime [12], cloaking at a distance [13], two-dimensional cloaks with arbitrary geometries [14], illusion optics [15],

DOI: 10.1201/9781003373117-4

an anti-cloak [16], scatterers and absorbers leading to super-scatterers [17] and super absorbers [18], beam shifters and splitters [19], beam benders/ expanders [20], reflectionless waveguide bends [21], expanding a narrow slit into a large window [22], polarization splitter and polarization rotator [23], and flat focusing lens [24], among others. In addition, the TO technique can be applied to the sources (e.g., current and charge distributions), using a process known as source transformation. This technique transforms the sources into a new current distribution while maintaining the same behavior. Proper material compensation is also included using the same transformation. The source transformation was first explored by Luo et al. [25].

This chapter aims to provide the reader with a comprehensive understanding of the TO approach, including the underlying concepts and derivations of the form-invariance of Maxwell's equations under coordinate transformations. In addition, the technique of source transformations is explored as a means of designing a linear antenna array. In this approach, each individual antenna element is transformed from a single dipole element in a free space, using a coordinate transformation and embedded in a complex electromagnetic media prescribed by the transformation designed by the TO technique. The performance of the proposed TO-based antenna array for phased array scanning is demonstrated through finite element method based on full-wave simulations using COMSOL Multiphysics ® [26]. This proposed antenna array holds promise for future applications in structurally integrated and conformal phased array antennas for wireless communications, radars, and sensing where antenna performance is influenced by structural and mechanical constraints.

4.2 FORM-INVARIANCE OF MAXWELL'S EQUATIONS AND ITS RELEVANCE TO TO

TO is a newly introduced technique that offers exceptional flexibility in designing electromagnetic devices through a coordinated transformation method. This technique also provides an excellent opportunity to observe and record novel wave-material interactions. The TO design approach is based on the fundamental assumption of the form-invariance of Maxwell's equations under coordinate transformations [6, 7]. In addition, it has been observed that the material parameters (ε, μ) in the transformed coordinate system can be interpreted as a set of material parameters in the original coordinate system [27]. Let's now consider the time-domain Maxwell's curl equations:

$$\nabla \times E = -j\omega\mu H \tag{4.1}$$

$$\nabla \times H = j\omega\mu E \tag{4.2}$$

As per the time-domain Maxwell's curl equations, E and H denote electric and magnetic fields, respectively, in a particular coordinate system, including Cartesian, cylindrical, or spherical coordinates. The divergence equations in the time domain are as follows:

$$B = \mu H \tag{4.3}$$

$$D = \mu E \tag{4.4}$$

where B and D are magnetic and electric flux densities, respectively and μ and ε are absolute permeability and absolute permittivity of the associated medium or material, respectively.

Next, let us consider a Cartesian coordinate system $G(x, y, z)$ to represent the original space and $G'(x', y', z')$ to represent the transformed space, as illustrated in Figure 4.1. The transformation from G to G' can be expressed as follows:

$$x' = x'(x, y, z)$$

$$y' = y'(x, y, z) \tag{4.5}$$

$$z' = z'(x, y, z)$$

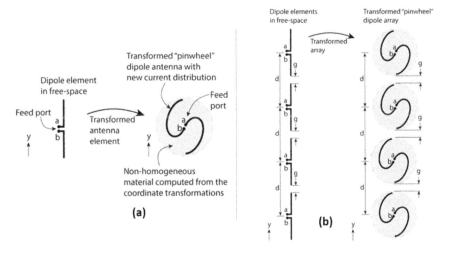

Figure 4.1 A depiction of transformation optics (TO) technique, where an original space, **G**, is transformed into a new space, **G'**, with the new material parameters ε, μ [29]. (a) current distribution in transformed 'pinwheel' dipole antenna; (b) current distribution in transformed 'pinwheel' dipole antenna array.

[Reprinted with permission from Eleftheriades et al. [29] © The IEEE]

Under this newly transformed coordinate system, Maxwell's equations remain form-invariant as follows [6, 7]:

$$\nabla \times E' = -j\omega\mu'H' \tag{4.6}$$

$$\nabla \times H' = j\omega\mu'E' \tag{4.7}$$

The material parameters in the transformed coordinate system are as follows [27]:

$$\varepsilon' = \frac{JJ_T}{\det J}\varepsilon \tag{4.8}$$

$$\mu' = \frac{JJT}{\det J}\mu \tag{4.9}$$

where J is the Jacobian matrix of the transformation from the $G(x,y,z)$ coordinate system to the new coordinate system, $G'(x',y',z')$, and J^T is the transpose of the matrix, J. J is defined as [28]:

$$J = \begin{vmatrix} \dfrac{\partial x'}{\partial x} & \dfrac{\partial x'}{\partial y} & \dfrac{\partial x'}{\partial z} \\[2mm] \dfrac{\partial y'}{\partial x} & \dfrac{\partial y'}{\partial y} & \dfrac{\partial y'}{\partial z} \\[2mm] \dfrac{\partial z'}{\partial x} & \dfrac{\partial z'}{\partial y} & \dfrac{\partial z'}{\partial z} \end{vmatrix} \tag{4.10}$$

The TO technique can be implemented following the steps listed next [28] (see Figure 4.1):

   I. Start with a known wave-material relation in the original coordinate system, such as a plane wave or a propagating Gaussian beam.
  II. Identify the volume of space in the original coordinate system and the corresponding volume of space in the transformed coordinate system.
 III. Determine the coordinate transformation needed to map the original space to the transformed space.
  IV. Calculate the material parameters in the transformed space using Equations 4.8 and 4.9.
   V. Transform the material parameters back to the original coordinate system to obtain the desired material.

Maxwell's equations are a fundamental set of partial differential equations in classical electromagnetism that describe the generation of electric and magnetic fields by charges and currents and how they interact. They are named after the physicist and mathematician James C. Maxwell, who used them to explain and propose that light is an electromagnetic phenomenon. It is well known in the electromagnetics community that Maxwell's equations are form-invariant [6, 7], meaning that they can be expressed and written in many different forms. The differential form of Maxwell's equations is as follows:

$$\nabla \times \bar{H} = \frac{\partial \bar{D}}{\partial t} + \bar{J} \qquad (4.11a)$$

$$\nabla \times \bar{E} = -\frac{\partial \bar{B}}{\partial t} \qquad (4.11b)$$

$$\nabla . \bar{D} = \sigma \qquad (4.11c)$$

$$\nabla . \bar{B} = 0 \qquad (4.11d)$$

where J is the electric current density and σ is the charge density. These equations can be rewritten in Cartesian components in the covariant notation, as follows:

$$\sum_j \sum_k \epsilon^{ijk} \frac{\partial H_k}{\partial x^j} = \frac{\partial D^i}{\partial t} + J^i \qquad (4.12a)$$

$$\sum_j \sum_k \epsilon^{ijk} \frac{\partial E_k}{\partial x^j} = -\frac{\partial B^i}{\partial t} \qquad (4.12b)$$

$$\sum_i \frac{\partial D^i}{\partial x^i} = \sigma \qquad (4.12c)$$

$$\sum_i \frac{\partial B^i}{\partial x^i} = 0 \qquad (4.12d)$$

where ϵ^{ijk} is a completely anti-symmetric Levi-Cevita tensor [30]. The Levi-Cevita tensor is defined as the following in Cartesian components:

$$\epsilon^{ijk} = \begin{cases} +1, \, (i,j,k) \Rightarrow (1,2,3), \, (2,3,1), \text{ or } (3,1,2) \\ -1, \, (i,j,k) \Rightarrow (3,2,1), \, (1,3,2), \text{ or } (2,1,3) \\ 0, \, i \Rightarrow j, \text{ or } j \Rightarrow k, \text{ or } k \Rightarrow i \end{cases} \qquad (4.13)$$

That is, ϵ^{ijk} is 1 if (i,j,k) results in an even permutation of $(1, 2, 3)$, -1 if the permutation is odd, and 0 in the case of index repetition.

The TO technique relies on the key assumption that the differential form of Maxwell's equations is form-invariant under any coordinate transformation, a well-known property of these equations [7]. As a result, during the transformation process, the associated fields are not affected, but the material parameters will be altered according to the new coordinate system or the transformed coordinate system. This property makes TO a powerful tool for designing electromagnetic and optical media with arbitrary shapes and materials that adopt wave propagation properties. Maxwell's equations can be expressed in the covariant notation using Cartesian components as follows:

$$\in^{ijk} \partial_j H_k = \varepsilon^{ij} \frac{\partial E_j}{\partial t} + J^i \tag{4.14a}$$

$$\in^{ijk} \partial_j E_k = -\mu^{ij} \frac{\partial H_j}{\partial t} \tag{4.14b}$$

$$\partial_j \varepsilon^{ij} E_j = \sigma \tag{4.14c}$$

$$\partial_j \mu^{ij} H_j = 0 \tag{4.14d}$$

Here, the indices (i, j, k) each run from 1 to 3, and x^i identifies a particular coordinate (i.e., x, y, and z), which means $x^1 => x$, $x^2 => y$, and $x^3 => z$.

To demonstrate the form-invariance of Maxwell's equations under coordinate transformations, we can consider the covariant form of the equations in Cartesian components (Equation 4.14a–d). We investigate what happens when we apply a coordinate transformation, $x' \rightarrow x$, where x is the original coordinate system and x' is the transformed coordinate system. We aim to show that Maxwell's equations will remain the same or form-invariant under this new coordinate system. To do this, we need to transform the electric field $(\bar{E})$ and magnetic field $(\bar{H})$ as follows:

$$E_{i'} = A_{i'}^i E_i$$
$$E_i = A_i^{i'} E_{i'}$$

and $H_i = A_i^{i'} H_{i'}$, where the Jacobian matrix is given by

$$A_i^{i'} = \frac{\partial x^{i'}}{\partial x^i}$$

The variable x^i represents the original coordinates, while $x^{i'}$ represents the transformed coordinate system. Applying the chain rule, it follows that the partial derivatives must also undergo a similar transformation:

$$\partial_{i'} = \frac{\partial}{\partial x^{i'}} = \left(\frac{\partial x^i}{\partial x^{i'}} \right) \frac{\partial}{\partial x^i}$$

and

$$\partial_{i'} = A_i^{i'} \partial_i \tag{4.15}$$

Now, substituting Equation 4.15) by Equation 4.14b, we get the following:

$$\in^{ijk} \partial_j \left(A_k^{k'} E_{k'} \right) = -\mu^{ij} \partial \left(\frac{A_j^{j'} H_{j'}}{\partial t} \right) \tag{4.16}$$

The left-hand side of Equation 4.16 can be written as the following:

$$\in^{ijk} \partial_j \left(A_k^{k'} E_{k'} \right) = \in^{ijk} A_k^{k'} \partial_j E_{k'} + \in^{ijk} E_{k'} \partial_j A_k^{k'}$$

$$\text{as } \frac{\partial}{\partial x}(u.v) = u\frac{\partial v}{\partial x} + v\frac{\partial u}{\partial x}$$

Note that, $\in^{ijk} \partial_j A_k^{k'} = \in^{ijk} \partial_j \partial_k x^{k'}$ as $A_i^{i'} = \dfrac{\partial x^{i'}}{\partial x^i}$.

Now rewriting the left-hand side of Equation 4.16, we get the following:

$$\in^{ijk} A_k^{k'} \partial_j E_{k'} + \in^{ijk} E_{k'} \partial_j \partial_k x^{k'}$$

$$= \in^{ijk} \left[A_k^{k'} \partial_j E_{k'} + E_{k'} \partial_k \partial_j x^{k'} \right]$$

$$= \in^{ijk} A_k^{k'} \partial_j E_{k'} \quad \left[\partial_j x^{k'} = \frac{\partial}{\partial x^j}\left(x^{k'} \right) = 0 \right]$$

$$= \in^{ijk} A_k^{k'} \frac{\partial x^{j'}}{\partial x^j} \frac{\partial}{\partial x^{j'}} E_{k'} \quad \left[\partial_j = \frac{\partial}{\partial x^j} = \left(\frac{\partial x^{j'}}{\partial x^j} \right) \cdot \frac{\partial}{\partial x^{j'}} \right]$$

$$= \in^{ijk} A_k^{k'} A_j^{j'} \partial_{j'} E_{k'}.$$

Therefore, Equation 4.16 can be written as the following:

$$\in^{ijk} A_k^{k'} A_j^{j'} \partial_{j'} E_{k'} = -\mu^{ij} \partial \left(\frac{A_j^{j'} H_{j'}}{\partial t} \right) \tag{4.17}$$

Now, using the definition of the determinant, we get the following:

$$\in^{i'j'k'} \left| A_n^{n'} \right| = A_i^{i'} A_k^{k'} A_j^{j'} \in^{ijk} \tag{4.18}$$

By inserting Equation 4.18 into Equation 4.17, we get the following:

$$\frac{\epsilon^{i'j'k'}\,|A|\partial_{j'}}{A_i^{i'}}\,E_{k'} = -\mu^{ij}A_j^{i'}\left(\frac{\partial H_{j'}}{\partial t}\right)$$

and

$$\epsilon^{i'j'k'}\,\partial_{j'}E_{k'} = -\frac{A_i^{i'}A_j^{j'}}{|A|}\mu^{ij}\left(\frac{\partial H_{j'}}{\partial t}\right) \tag{4.19}$$

Again, substituting Equation 4.15 by Equation 4. 14a, we get the following:

$$\epsilon^{ijk}\,\partial_j\left(A_k^{k'}H_{k'}\right) = \varepsilon^{ij}\,\partial\left(\frac{A_j^{j'}E_{j'}}{\partial t}\right) + J^i \tag{4.20}$$

The left-hand side of Equation 4.20 can be written as the following:

$$\epsilon^{ijk}\,\partial_j\left(A_k^{k'}H_{k'}\right) = \epsilon^{ijk}\,A_k^{k'}\partial_j H_{k'} + \epsilon^{ijk}\,H_{k'}\partial_j A_k^{k'}$$

$$\text{as}\,\frac{\partial}{\partial x}(u.v) = u\frac{\partial v}{\partial x} + v\frac{\partial u}{\partial x}$$

Note that, $\epsilon^{ijk}\,\partial_j A_k^{k'} = \epsilon^{ijk}\,\partial_j\partial_k x^{k'}$ as $A_i^{i'} = \dfrac{\partial x^{i'}}{\partial x^i}$

Now rewriting the left-hand side of Equation 4.20:

$$\epsilon^{ijk}\,A_k^{k'}\partial_j H_{k'} + \epsilon^{ijk}\,H_{k'}\partial_j\partial_k x^{k'}$$

$$= \epsilon^{ijk}\left[A_k^{k'}\partial_j E_{k'} + H_{k'}\partial_k\partial_j x^{k'}\right]$$

$$= \epsilon^{ijk}\,A_k^{k'}\partial_j H_{k'}\left[\partial_j x^{k'} = \frac{\partial}{\partial x^j}(x^{k'}) = 0\right]$$

$$= \epsilon^{ijk}\,A_k^{k'}\frac{\partial x^{i'}}{\partial x^j}\frac{\partial}{\partial x^{i'}}H_{k'}\left[\partial_j = \frac{\partial}{\partial x^j} = \left(\frac{\partial x^{i'}}{\partial x^j}\right).\frac{\partial}{\partial x^{i'}}\right]$$

$$= \epsilon^{ijk}\,A_k^{k'}A_j^{j'}\partial_{j'}H_{k'}$$

Therefore, Equation 4.20 can be written as the following:

$$\epsilon^{ijk}\,A_k^{k'}A_j^{j'}\partial_{j'}H_{k'} = \varepsilon^{ij}\,\partial\left(\frac{A_j^{j'}E_{j'}}{\partial t}\right) + J^i \tag{4.21}$$

Now, using the definition of the determinant, we get the following:

$$\epsilon^{i'j'k'} \left| A^{n'}_n \right| = A^{i'}_i A^{k'}_k A^{j'}_j \, \epsilon^{ijk} \tag{4.22}$$

By inserting Equation 4.22 into Equation 4.21, we get the following:

$$\frac{\epsilon^{i'j'k'} |A| \partial_{j'}}{A^{i'}_i} H_{k'} = \varepsilon^{ij} \partial \left(\frac{A^{j'}_j E_{j'}}{\partial t} \right) + J^i$$

$$\epsilon^{i'j'k'} \partial_{j'} H_{k'} = \frac{A^{i'}_i A^{j'}_j}{|A|} \varepsilon^{ij} \left(\frac{\partial E_{j'}}{\partial t} \right) + \frac{A^{i'}_i}{|A|} J^i \tag{4.23}$$

Now, Gauss's law from Equation 4.14c can go under transformation and rewriting as the following:

$$\partial_i \varepsilon^{ij} E_j = \sigma$$

$$A^{i'}_i \partial_{i'} \varepsilon^{ij} A^{j'}_j E_{j'} = \sigma$$

$$\frac{A^{i'}_i \partial_{i'} \varepsilon^{ij} A^{j'}_j E_{j'}}{|A|} = \frac{\sigma}{|A|} \tag{4.24}$$

Rewriting the left-hand side of Equation 4.24, we get the following:

$$\partial_{i'} \frac{A^{i'}_i \varepsilon^{ij} A^{j'}_j E_{j'}}{|A|} = A^{i'}_i \varepsilon^{ij} E_{j'} \partial_{i'} \frac{A^{i'}_i}{|A|} + \frac{A^{j'}_j \partial_{i'} \left(A^{i'}_i \varepsilon^{ij} E_{j'} \right)}{|A|}$$

Hence, Equation 4.24 can be rewritten as the following:

$$\partial_{i'} \frac{A^{i'}_i \varepsilon^{ij} A^{j'}_j E_{j'}}{|A|} - A^{i'}_i \varepsilon^{ij} E_{j'} \partial_{i'} \frac{A^{i'}_i}{|A|} = \frac{\sigma}{|A|} \tag{4.25}$$

Based on the derivation presented by Mitra [31], we can state that the second term on the left-hand side of Equation 4.25 is equal to 0. Furthermore, the inverse of the matrix inverse can be obtained using the co-factor formula:

$$\frac{A^{i'}_i}{\left| A^{k'}_k \right|} = \frac{\epsilon^{i'u'v'} \in_{iuv} A^{u'}_u A^{v'}_v}{2} \tag{4.26}$$

and taking the partial derivative:

$$\frac{\partial_{i'} A^{u'}_u \epsilon^{i'u'v'} \in_{iuv} A^{u'}_u A^{v'}_v}{2} \tag{4.27}$$

According to the derivations from Mitra [31], this term is also zero and as the partial derivatives are commutative, i.e., $\dfrac{\partial}{\partial x^i} \cdot \dfrac{\partial}{\partial x^u} = \dfrac{\partial}{\partial x^u} \cdot \dfrac{\partial}{\partial x^i}$, while the exchange of i' and u' is anti-symmetric. Rewriting Equation 4.25 results the following:

$$\partial_{i'} \frac{A_i^{i'} \varepsilon^{ij} A_j^{j'} E_{j'}}{|A|} = \frac{\sigma}{|A|} \tag{4.28}$$

Similarly, Equation 4.14d can be transformed and written as follows:

$$\partial_i \mu^{ij} H_j = 0,$$

$$A_i^{i'} \partial_{i'} \mu^{ij} A_j^{j'} H_{j'} = 0,$$

and

$$\frac{A_i^{i'} \partial_{i'} \mu^{ij} A_j^{j'} H_{j'}}{|A|} = 0. \tag{4.29}$$

Rewriting the left-hand side of Equation 4.29, we get the following:

$$\partial_{i'} \frac{A_i^{i'} \mu^{ij} A_j^{j'} H_{j'}}{|A|} = A_i^{i'} \mu^{ij} H_{j'} \partial_{i'} \frac{A_i^{i'}}{|A|} + \frac{A_i^{i'} \partial_{i'} \left(A_i^{i'} \mu^{ij} H_{j'} \right)}{|A|} \tag{4.30}$$

Rewriting the left-hand side of Equation 4.30, we get the following:

$$\partial_{i'} \frac{A_i^{i'} \mu^{ij} A_j^{j'} H_{j'}}{|A|} = A_i^{i'} \mu^{ij} H_{j'} \partial_{i'} \frac{A_i^{i'}}{|A|} + \frac{A_i^{i'} \partial_{i'} \left(A_i^{i'} \mu^{ij} H_{j'} \right)}{|A|}$$

Hence, rewriting Equation 4.29, we get the following:

$$\partial_{i'} \frac{A_i^{i'} \mu^{ij} A_j^{j'} H_{j'}}{|A|} - A_i^{i'} \mu^{ij} H_{j'} \partial_{i'} \frac{A_i^{i'}}{|A|} = 0 \tag{4.31}$$

According to Mitra [31], the second term on the left-hand side of Equation 4.31 is 0, and we can calculate the inverse of the matrix inverse using the co-factor formula:

$$\frac{A_i^{i'}}{|A_k^{k'}|} = \frac{\epsilon^{i'u'v'} \epsilon_{iuv} A_u^{u'} A_v^{v'}}{2} \tag{4.32}$$

Table 4.1 The Form-Invariance of Maxwell's Equations under a Coordinate Transformation and the Corresponding Transformation Rules for Material Parameters, Charges, and Current Distribution in the New Coordinates Have Been Extensively Studied in the Literature [33, 34].

Original equations	Equations after transformation	Rules for transformation	Transformed equations using the rules
$\epsilon^{ijk}\,\partial_j H_k = \varepsilon^{ij}\dfrac{\partial E_j}{\partial t} + J^i$	$\epsilon^{i'j'k'}\,\partial_{j'} H_{k'} = \dfrac{A_i^{i'}A_j^{j'}}{\lvert A\rvert}\varepsilon^{ij}\left(\dfrac{\partial E_j}{\partial t}\right) + \dfrac{A_i^{i'}}{\lvert A\rvert}J^i$	$\varepsilon^{i'j'} = \dfrac{A_i^{i'}A_j^{j'}}{\lvert A\rvert}\varepsilon^{ij}$	$\epsilon^{i'j'k'}\,\partial_{j'} H_{k'} = \varepsilon^{i'j'}\left(\dfrac{\partial E_{j'}}{\partial t}\right) + J^{i'}$
$\epsilon^{ijk}\,\partial_j E_k = -\mu^{ij}\dfrac{\partial H_j}{\partial t}$	$\epsilon^{i'j'k'}\,\partial_{j'} E_{k'} = -\dfrac{A_i^{i'}A_j^{j'}}{\lvert A\rvert}\mu^{ij}\left(\dfrac{\partial H_j}{\partial t}\right)$	$\mu^{i'j'} = \dfrac{A_i^{i'}A_j^{j'}}{\lvert A\rvert}\mu^{ij}$	$\epsilon^{i'j'k'}\,\partial_{j'} E_{k'} = -\mu^{i'j'}\left(\dfrac{\partial H_{j'}}{\partial t}\right)$
$\partial_i \varepsilon^{ij} E_j = \sigma$	$\partial_{i'}\dfrac{A_i^{i'}\varepsilon^{ij}A_j^{j'}E_j}{\lvert A\rvert} = \dfrac{\sigma}{\lvert A\rvert}$	$\rho' = \dfrac{\sigma}{\lvert A\rvert}$	$\partial_{i'}\varepsilon^{i'j'}E_{j'} = \sigma'$
$\partial_i \mu^{ij} H_j = 0$	$\partial_{i'}\dfrac{A_i^{i'}\mu^{ij}A_j^{j'}H_j}{\lvert A\rvert} = 0$	$J^{i'} = \dfrac{A_i^{i'}}{\lvert A\rvert}J^i$	$\partial_{i'}\mu^{i'j'}H_{j'} = 0$

and taking the partial derivative:

$$\frac{\partial_{i'} A_u^{u'} \epsilon^{i'u'v'} \epsilon_{iuv} A_u^{u'} A_v^{v'}}{2} \tag{4.33}$$

Rewriting Equation 4.31 results in the following:

$$\partial_{i'} \frac{A_i^{i'} \mu^{ij} A_j^{j'} H_{j'}}{|A|} = 0 \tag{4.34}$$

All the equations demonstrate how the material parameters, current sources, and charges can be expressed in the transformed space through the appropriate coordinate transformation rules while maintaining the form-invariance of Maxwell's equations in the new coordinate system. The process outlined in these derivations follows closely the works of Mitra, Kundtz, Johnson, and Allen [31–34] regarding the form-invariance of Maxwell's equations under the coordinate system of transformations.

4.3 DESIGN OF PHASED ARRAY ANTENNA ELEMENTS USING TO

In this section, the use of TO is extended to an array of pinwheel-shaped elements, building upon the concept of TO for individual elements. The efficacy of this approach is illustrated through full-wave finite element analysis, which demonstrates that a transformed "pinwheel" linear array can function similarly to a uniformly spaced linear dipole array, preserving all the benefits of array processing. This same method can also be employed to transform a region with sources, such as current and charge distributions, so that they behave similarly to the untransformed sources. By utilizing source transformations, intricate radiation patterns can be designed with engineered material properties, allowing transformed geometries to reproduce the performance of the original structures. This is particularly advantageous when physical constraints necessitate modifications to the spatial configuration of the source. Further insight into the application of source transformations in antenna design can be found in Luo et al., Kundtz et al., and Allen et al. [25, 35, 36]. A visualization of the TO technique is illustrated in Figure 4.2.

Luo et al. [25] were among the first to propose an application of the TO technique by transforming a dipole current source into a new distribution while maintaining its characteristics as a dipole antenna. Kundtz et al. [35] extended this concept by introducing an optical source transformation with a "pinwheel" transformation. They approximated the sheet current of a simple dipole as a volume current and transformed it into a new current distribution using a "pinwheel" coordinate transformation.

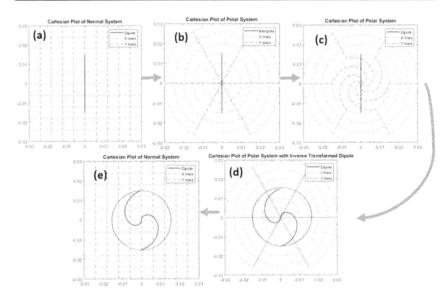

Figure 4.2 The TO technique has been utilized to design a material-embedded antenna array, as described by Mitra and Mitra et al. [31, 37]. This involves transforming a single dipole antenna element into a "pinwheel" shape using TO-embedded media, as shown in Figure (4.2a), and then transforming a linear dipole array (reference array) into a linear array of "pinwheel" coordinate transformation, as mentioned in Figure 4.2c, 4.2d, 4.2e.

In this design, the source transformation technique is utilized to create a new linear array in which each antenna element is transformed from a single dipole element in free space (as illustrated in Figure 4.3a) [31, 37]. The resulting antenna array radiates the same field as a linear dipole array, as shown in Figure 4.3b. The individual antenna element in the transformed linear array, as demonstrated in Figure 4.3a, is a pinwheel-shaped antenna that is an extreme example of the transformation of a dipole antenna element similar to the transformation presented by Kundtz et al. [35]. In addition, the transformed antenna element in the array is encompassed by a complex electromagnetic medium prescribed by the transformation, as depicted by the dotted gray circle in Figure 4.3b.

In the following, let us examine the N-element dipole phased array along the y-axis presented in Figure 4.3b (left), which is positioned in free space and referred to as the "reference array." The elements in the "reference array" are equally spaced with an edge-to-edge distance of g = $\lambda/15$, where λ is the wavelength of the electromagnetic wave in free space in which the phased array is intended to operate. For illustrating the proposed "pinwheel" array, we consider a two-dimensional (2D) space.

The current distribution for each element in the "reference array" is represented by $I_n = J.e^{i(n-1)\emptyset}$, where J approximates the current distribution on a

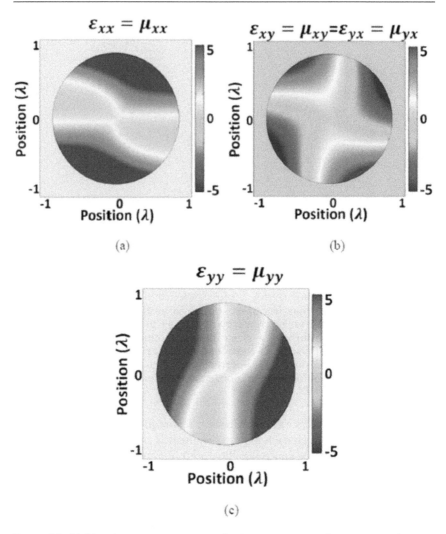

Figure 4.3 (a) New linear array is created using source transformation technique. (b) and (c) The resulting antenna array radiates the same field as a linear dipole array.

thin wire at $x = 0$ and $n = 1, 2 \ldots$ N, with N being the number of elements and $\emptyset$ being the phase difference between the adjacent elements. The proposed method is demonstrated using a four-element array, where each dipole is $\lambda/2$ in length and spans 4.2.2λ. The aim is to transform the "reference array" into a linear array of equidistant complex geometry antennas, with each element taking the form of a "pinwheel," as shown in Figure 4.3b. The first step involves using the source transformation method to transform each dipole into a "pinwheel" antenna element, which is depicted in Figure 4.3c.

Table 4.1 gives the material parameters for the space under coordinate transformation, where

$$\mu' = \frac{\varepsilon'}{} = \frac{A\varepsilon A^T}{\det A} \tag{4.35}$$

$A = \partial(x',y',z')/\partial(x,y,z)$ is the Jacobian matrix, and A^T represents the transpose of the Jacobian matrix. The coordinate transformation from Cartesian to cylindrical coordinates is given by the following:

$$\rho = \sqrt{x^2 + y^2}$$

$$\theta = tan^{-1}\left(\frac{y}{x}\right) \tag{4.36}$$

$$z = z$$

The following equation gives the Jacobian matrix of the transformation:

$$A1 = \begin{bmatrix} \cos(\theta) & \sin(\theta) & 0 \\ -\dfrac{\sin(\theta)}{\rho} & \dfrac{\cos(\theta)}{\rho} & 0 \\ 0 & 0 & 1 \end{bmatrix} \tag{4.37}$$

where $|A1| = \dfrac{1}{\rho}$

The following equation gives the derivation of the material parameters:

$$\varepsilon' = \mu' = \frac{A1 * \varepsilon * A1^T}{|A1|} = \begin{bmatrix} \rho & 0 & 0 \\ 0 & \dfrac{1}{\rho} & 0 \\ 0 & 0 & \rho \end{bmatrix} \tag{4.38}$$

The following gives the transformation from cylindrical to "pinwheel" coordinates transformation:

$$\rho' = \rho$$

$$\theta' = \begin{cases} \theta & , \rho > R_1 \\ \theta + \Delta\theta\left(1 - \dfrac{\rho}{R_1}\right), & \rho < R_1, \end{cases} \tag{4.39}$$

and

$$z' = z$$

The Jacobian for this transformation is

$$A2 = \begin{bmatrix} \dfrac{\partial \rho'}{\partial \rho} & \dfrac{\partial \rho'}{\partial \theta} & \dfrac{\partial \rho'}{\partial z} \\[2mm] \dfrac{\partial \theta'}{\partial \rho} & \dfrac{\partial \theta'}{\partial \theta} & \dfrac{\partial \theta'}{\partial z} \\[2mm] \dfrac{\partial z'}{\partial \rho} & \dfrac{\partial z'}{\partial \theta} & \dfrac{\partial z'}{\partial z} \end{bmatrix} \tag{4.40}$$

Now,

$$\frac{\partial \rho'}{\partial \rho} = 1, \frac{\partial \rho'}{\partial \theta} = 0, \frac{\partial \rho'}{\partial z} = 0$$

$$\frac{\partial \theta'}{\partial \rho} = \frac{\partial}{\partial \rho}\left\{ \theta + \Delta\theta\left(1 - \frac{\rho}{R_1}\right) \right\} = \frac{\partial}{\partial \rho}\left(\theta + \Delta\theta - \frac{\Delta\theta\rho}{R_1} \right) = -\frac{\Delta\theta}{R_1}\left(\frac{\partial \rho}{\partial \rho} \right) = -\frac{\Delta\theta}{R_1}$$

$$\frac{\partial \theta'}{\partial \theta} = \frac{\partial}{\partial \theta}\left\{ \theta + \Delta\theta\left(1 - \frac{\rho}{R_1}\right) \right\} = \frac{\partial \theta}{\partial \theta} = 1$$

$$\frac{\partial \theta'}{\partial z} = 0, \frac{\partial z'}{\partial \rho} = 0, \frac{\partial z'}{\partial \theta} = 0, \text{ and } \frac{\partial z'}{\partial z} = 1$$

Equation 4.40 can be rewritten as the following:

$$A2 = \begin{bmatrix} 1 & 0 & 0 \\[2mm] -\dfrac{\Delta\theta}{R_1} & 1 & 0 \\[2mm] 0 & 0 & 1 \end{bmatrix} \tag{4.41}$$

where $|A2| = 1$.

Figure 4.4 The process of including sources using the "pinwheel" transformation in transformation electromagnetics is demonstrated here step-by-step. The source transformation technique involves the line current of a dipole antenna and is based on previous works by Luo et al. and Kundtz et al. [25, 35]. Figure 4.4a shows the definition of the current in a Cartesian coordinate system. In Figure 4.4b, the current is transformed into a cylindrical coordinate system, and in Figure 4.4c, it is further transformed into the "pinwheel" coordinates. In Figure 4.4d, the transformation is applied

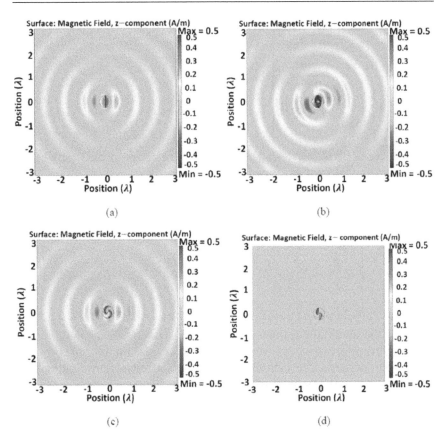

Figure 4.4 The z-component of the magnetic field of a single antenna element is shown by Mitra and Mitra et al. [31, 37], where (a) represents a dipole antenna of length L = λ/2 in free space; (b) represents a dipole that has undergone a "pinwheel" rotation of Δθ = 180° without any material compensation, as calculated using Equation 4.48; (c) represents a dipole that has undergone a "pinwheel" rotation of Δθ = 180° with proper material compensation, also calculated using Equation 4.48; and (d) represents the difference between the fields in (a) and (c).

to the dipole line current, and in Figure 4.4c, the resulting current distribution is expressed in the original coordinates. The circle in Figures 4.4d and 4.4e refers to the material shell. [Adapted with permission from Kundtz et al. [35] © The Optical Society]

The following equation gives the derivation of material parameters in the pinwheel coordinates:

$$\varepsilon'' = \mu'' = \frac{A2 * \varepsilon' * A2^T}{|A2|}$$

$$= \begin{bmatrix} \rho & -\dfrac{\rho\Delta\theta}{R_1} & 0 \\ -\dfrac{\rho\Delta\theta}{R_1} & \dfrac{1}{\rho}+\dfrac{\rho\Delta\theta^2}{R_1^{\,2}} & 0 \\ 0 & 0 & 1 \end{bmatrix} \qquad (4.42)$$

Adapting from Equation 4.42, the material parameters in the "pinwheel" coordinates can be written as the following:

$$\varepsilon'' = \mu'' = \begin{bmatrix} \rho' & -\dfrac{\rho'\Delta\theta}{R_1} & 0 \\ -\dfrac{\rho'\Delta\theta}{R_1} & \dfrac{1}{\rho'}+\dfrac{\rho'\Delta\theta^2}{R_1^{\,2}} & 0 \\ 0 & 0 & 1 \end{bmatrix} \qquad (4.43)$$

To obtain the material parameters in the Cartesian coordinate system, the cylindrical transformation is inverted. This allows us to establish the inverse relationship between the cylindrical and the Cartesian coordinates and retrieve the necessary material parameters.

The inverse relations are: $x = \rho\cos\theta, y = \rho\sin\theta$, and $z = z$.

The Jacobian for this inverse transformation is

$$A3 = \begin{bmatrix} \cos\theta & -\rho\sin\theta & 0 \\ \sin\theta & \rho\cos\theta & 0 \\ 0 & 0 & 1 \end{bmatrix} \qquad (4.44)$$

where $|A3| = \rho$.

"Dropping the primes," the following equation gives the new form of Equation 4.43):

$$\varepsilon'' = \mu'' = \begin{bmatrix} \rho & -\dfrac{\rho\Delta\theta}{R_1} & 0 \\ -\dfrac{\rho\Delta\theta}{R_1} & \dfrac{1}{\rho}+\dfrac{\rho\Delta\theta^2}{R_1^{\,2}} & 0 \\ 0 & 0 & 1 \end{bmatrix} \qquad (4.45)$$

The material parameters can be derived as the following:

$$\varepsilon''' = \mu''' = \frac{A3 * \varepsilon'' * A3^T}{|A3|}$$

$$= \frac{1}{\rho} * \begin{bmatrix} \left(x+\dfrac{\rho\Delta\theta y}{R_1}\right) & -\left\{\dfrac{\Delta\theta x}{R_1}+y\left(\dfrac{1}{\rho}+\dfrac{\rho\Delta\theta^2}{R_1^{\,2}}\right)\right\} & 0 \\[2ex] \left(y-\dfrac{\rho\Delta\theta x}{R_1}\right) & -\left\{\dfrac{\Delta\theta y}{R_1}-x\left(\dfrac{1}{\rho}+\dfrac{\rho\Delta\theta^2}{R_1^{\,2}}\right)\right\} & 0 \\[2ex] 0 & 0 & 1 \end{bmatrix} * \begin{bmatrix} cos\theta & sin\theta & 0 \\ -\rho sin\theta & \rho cos\theta & 0 \\ 0 & 0 & 1 \end{bmatrix}$$

$$= \frac{1}{\rho} * \begin{bmatrix} A & B & 0 \\ C & D & 0 \\ 0 & 0 & 1 \end{bmatrix} * \begin{bmatrix} cos\theta & sin\theta & 0 \\ -\rho sin\theta & \rho cos\theta & 0 \\ 0 & 0 & 1 \end{bmatrix} \tag{4.46}$$

where inverse relations are in use and $A = \left(x+\dfrac{\rho\Delta\theta y}{R_1}\right)$, $B = -\left\{\dfrac{\Delta\theta x}{R_1}+\right.$

$\left. y\left(\dfrac{1}{\rho}+\dfrac{\rho\Delta\theta^2}{R_1^{\,2}}\right)\right\}$, $C = \left(y-\dfrac{\rho\Delta\theta x}{R_1}\right)$, and $D = -\left\{\dfrac{\Delta\theta y}{R_1}-x\left(\dfrac{1}{\rho}+\dfrac{\rho\Delta\theta^2}{R_1^{\,2}}\right)\right\}$.

Equation 4.46 can be rewritten as the following:

$$\varepsilon''' = \mu''' = \frac{1}{\rho} * \begin{bmatrix} Acos\theta - B\rho sin\theta & Asin\theta + B\rho cos\theta & 0 \\ Ccos\theta - D\rho sin\theta & Csin\theta + D\rho cos\theta & 0 \\ 0 & 0 & 1 \end{bmatrix}$$

$$\varepsilon''' = \mu''' = \begin{bmatrix} \left(\dfrac{Acos\theta}{\rho} - Bsin\theta\right) & \left(\dfrac{Asin\theta}{\rho} + Bcos\theta\right) & 0 \\[2ex] \left(\dfrac{Ccos\theta}{\rho} - Dsin\theta\right) & \left(\dfrac{Csin\theta}{\rho} + Dcos\theta\right) & 0 \\[2ex] 0 & 0 & 1/\rho \end{bmatrix} \tag{4.47}$$

Expanding on Equation 4.47, we can express the material parameters in Cartesian coordinates as follows:

$$\varepsilon''' = \mu''' = \begin{pmatrix} 1 + \dfrac{y*\Delta\theta\left(\dfrac{2R_1 x}{\rho} + y*\Delta\theta\right)}{R_1^2} & -\dfrac{\Delta\theta\left(xy\Delta\theta + R_1\rho\cos(2\theta)\right)}{R_1^2} & 0 \\[4mm] -\dfrac{\Delta\theta\left(xy\Delta\theta + R_1\rho\cos(2\theta)\right)}{R_1^2} & 1 + \dfrac{x*\Delta\theta\left(-\dfrac{2R_1 y}{\rho} + x*\Delta\theta\right)}{R_1^2} & 0 \\[4mm] 0 & 0 & 1 \end{pmatrix} \qquad (4.48)$$

Equation 4.48 leads to the realization of anisotropic and inhomogeneous permittivity and permeability tensors, where both electromagnetic parameters ε and μ exhibit similar behaviors as shown in Figure 4.4. Such a transformation medium can be achieved through discrete metamaterials and structures such as periodic split ring resonators (SRRs) [38].

Figure 4.5 Spatial variation of the material parameters inside the shell [31,37]: (a) $\varepsilon_{xx} = \mu_{xx}$, (b) $\varepsilon_{xy} = \mu_{xy} = \varepsilon_{yx} = \mu_{yx}$, (c) $\varepsilon_{yy} = \mu_{yy}$. The material parameters ε_{xy}, μ_{xy}, ε_{yx}, and μ_{yx} are equal.

As described by Kundtz et al. [35], the current distribution of a dipole along $x = 0$ in Figure 4.3.1a was chosen as the following:

$$\bar{j} = \begin{pmatrix} 0 \\ \dfrac{1}{\sqrt{\delta*\pi}}e^{-\frac{x^2}{\delta}} \\ 0 \end{pmatrix} = \begin{pmatrix} J_x \\ J_y \\ J_z \end{pmatrix}, \qquad (4.49)$$

The limit $\delta \rightarrow 0$ can be taken to approximate Equation 4.49 to the sheet current density, which is set to be infinitesimally small relative to the length of the dipole. This allows the half-wave dipole to be initially defined by its current distribution in Cartesian coordinates and then converted into cylindrical coordinates to facilitate the transformation.

Table 1 gives the transformed current density, J', under coordinate transformation:

$$J' = \frac{A}{|A|}J. \qquad (4.50)$$

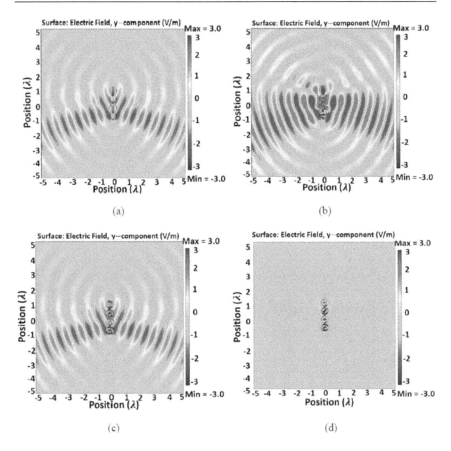

Figure 4.5 Total electric field distributions for three distinct array configurations for a scan angle of $_{,s} = 24.2.5$ degrees [31, 37]. These include (a) the reference or original linear dipole antenna array, (b) the "pinwheel" antenna array without any material compensation, (c) the material-embedded pinwheel-shaped antenna linear array, and (d) the difference between the electric fields in (a) and (c).

The coordinate transformation from Cartesian to cylindrical coordinates is defined by the following:

$$\rho = \sqrt{x^2 + y^2},$$

$$\theta = \arctan\left(\frac{y}{x}\right), \tag{4.51}$$

and

$$z = z.$$

The Jacobian of this transformation is

$$A1 = \begin{bmatrix} \cos(\theta) & \sin(\theta) & 0 \\ -\dfrac{\sin(\theta)}{\rho} & \dfrac{\cos(\theta)}{\rho} & 0 \\ 0 & 0 & 1 \end{bmatrix},$$

(4.52)

where $|A1| = \dfrac{1}{\rho}$.

The current can be written in the cylindrical coordinates as the following:

$$\overline{J}_1 = \frac{A_1}{|A_1|}\overline{J}.$$

(4.53)

Now,

$$\frac{A_1}{|A_1|} = \rho \begin{bmatrix} \cos(\theta) & \sin(\theta) & 0 \\ -\dfrac{\sin(\theta)}{\rho} & \dfrac{\cos(\theta)}{\rho} & 0 \\ 0 & 0 & 1 \end{bmatrix}$$

$$= \begin{bmatrix} \rho\cos(\theta) & \rho\sin(\theta) & 0 \\ -\sin\theta & \cos\theta & 0 \\ 0 & 0 & \rho \end{bmatrix}.$$

So, we can write Equation 4.53 as the following:

$$\overline{J}_1 = \begin{bmatrix} \rho\cos(\theta) & \rho\sin(\theta) & 0 \\ -\sin\theta & \cos\theta & 0 \\ 0 & 0 & \rho \end{bmatrix} * \begin{pmatrix} 0 \\ J_y \\ 0 \end{pmatrix},$$

$$= \begin{bmatrix} 0 + j_y\rho\sin\theta + 0 \\ 0 + j_y\cos\theta + 0 \\ 0 + 0 + 0 \end{bmatrix},$$

$$= J_y * \begin{bmatrix} \rho\sin\theta \\ \cos\theta \\ 0 \end{bmatrix}.$$

Also,

$$\frac{1}{\sqrt{\delta * \pi}} exp\left(-\frac{(\rho cos\theta)^2}{\delta}\right) * \begin{bmatrix} \rho sin\theta \\ cos\theta \\ 0 \end{bmatrix} = K * \begin{bmatrix} \rho sin\theta \\ cos\theta \\ 0 \end{bmatrix}, \tag{4.54}$$

where, $K = \dfrac{1}{\sqrt{\delta * \pi}} exp\left(-\dfrac{(\rho cos\theta)^2}{\delta}\right)$.

The Jacobian for the cylindrical to the "pinwheel" coordinate transformations is given by Equation 4.41 as the following:

$$A2 = \begin{bmatrix} 1 & 0 & 0 \\ -\dfrac{\Delta\theta}{R_1} & 1 & 0 \\ 0 & 0 & 1 \end{bmatrix}, |A2| = 1.$$

The current can be written in the "pinwheel" coordinates as the following:

$$\overline{J_2} = \frac{A_2}{|A_2|} \overline{J_1},$$

$$= \begin{bmatrix} 1 & 0 & 0 \\ -\dfrac{\Delta\theta}{R_1} & 1 & 0 \\ 0 & 0 & 1 \end{bmatrix} * \begin{bmatrix} K\rho sin\theta \\ Kcos\theta \\ 0 \end{bmatrix}$$

$$= \frac{K}{R_1} \begin{bmatrix} R_1\rho sin\theta \\ R_1 cos\theta - \rho\Delta\theta sin\theta \\ 0 \end{bmatrix}.$$

Then,

$$\overline{J_2} = \frac{1}{R_1\sqrt{\delta * \pi}} exp\left(-\frac{(\rho cos\theta)^2}{\delta}\right) * \begin{bmatrix} R_1\rho sin\theta \\ R_1 cos\theta - \rho\Delta\theta sin\theta \\ 0 \end{bmatrix}. \tag{4.55}$$

In "pinwheel" coordinates:

$$\rho' = \rho;$$

for $\rho < R_1$,

$$\theta' = \theta + \Delta\theta\left(1 - \frac{\rho'}{R_1}\right);$$

and

$$\theta = \theta' - \Delta\theta\left(1 - \frac{\rho'}{R_1}\right) = -\left\{\Delta\theta\left(1 - \frac{\rho'}{R_1}\right) - \theta'\right\} = -u'.$$

Substituting this by Equation 4.55 gives

$$
\overline{J_2} = \frac{1}{R_1\sqrt{\delta}*\pi}\,exp\left(-\frac{\left(\rho'\cos(-u')\right)^2}{\delta}\right)*\begin{bmatrix} R_1\rho'\sin(-u') \\ R_1\cos(-u') - \rho'\Delta\theta\sin(-u') \\ 0 \end{bmatrix}
$$

$$
= \frac{1}{R_1\sqrt{\delta}*\pi}\,exp\left(-\frac{\left(\rho'\cos(u')\right)^2}{\delta}\right)*\begin{bmatrix} -R_1\rho'\sin(u') \\ R_1\cos(u') + \rho'\Delta\theta\sin(u') \\ 0 \end{bmatrix}, \qquad (4.56)
$$

where, $u' = \left\{\Delta\theta\left(1 - \frac{\rho'}{R_1}\right) - \theta'\right\}$.

To obtain the current distribution in Cartesian coordinates, it is necessary to perform the inverse transformation from cylindrical coordinates. The inverse relations are given by the following:

$$x = \rho\cos\theta, \ y = \rho\sin\theta, \text{ and } z = z.$$

The following equation gives the Jacobian of these inverse relations:

$$
A3 = \begin{bmatrix} \cos\theta & -\rho\sin\theta & 0 \\ \sin\theta & \rho\cos\theta & 0 \\ 0 & 0 & 1 \end{bmatrix},
$$

where $|A3| = \rho$.

After "dropping the primes," the current distribution from Equation 4.56 can be rewritten as the following:

$$\overline{J}_2 = \frac{1}{R_1\sqrt{\delta}*\pi} exp\left(-\frac{\left(\rho\cos(u)\right)^2}{\delta}\right)*\begin{bmatrix} -R_1\rho\sin(u) \\ R_1\cos(u)+\rho\Delta\theta\sin(u) \\ 0 \end{bmatrix},$$

$$= B*\begin{bmatrix} -R_1\rho\sin(u) \\ R_1\cos(u)+\rho\Delta\theta\sin(u) \\ 0 \end{bmatrix}, \qquad (4.57)$$

where, $B = \dfrac{1}{R_1\sqrt{\delta}*\pi} exp\left(-\dfrac{\left(\rho\cos(u)\right)^2}{\delta}\right)$.

The following gives the derived currents:

$$\overline{J}_3 = \frac{A_3}{|A_3|}\overline{J}_2,$$

$$= B*\begin{bmatrix} \cos\theta/\rho & -\sin\theta & 0 \\ \sin\theta/\rho & \cos\theta & 0 \\ 0 & 0 & \frac{1}{\rho} \end{bmatrix}*\begin{bmatrix} -R_1\rho\sin(u) \\ R_1\cos(u)+\rho\Delta\theta\sin(u) \\ 0 \end{bmatrix}$$

$$= B*\begin{bmatrix} \left(\frac{\cos\theta}{\rho}\right)(-R_1\rho\sin(u))+(-\sin\theta)(R_1\cos(u)+\rho\Delta\theta\sin(u))+0 \\ (\sin\theta/\rho)(-R_1\rho\sin(u))+(\cos\theta)(R_1\cos(u)+\rho\Delta\theta\sin(u))+0 \\ 0+0+0 \end{bmatrix}$$

$$= B*\begin{bmatrix} \alpha \\ \beta \\ 0 \end{bmatrix} \qquad (4.58)$$

where, $\alpha = \left(\dfrac{\cos\theta}{\rho}\right)(-R_1\rho\sin(u))+(-\sin\theta)(R_1\cos(u)+\rho\Delta\theta\sin(u))$ and

$\beta = (\sin\theta/\rho)(-R_1\rho\sin(u))+(\cos\theta)(R_1\cos(u)+\rho\Delta\theta\sin(u))$.

Now,

$$\alpha = \left(\frac{cos\theta}{\rho}\right)\left(-R_1\rho\sin(u)\right) + \left(-sin\theta\right)\left(R_1\cos(u) + \rho\Delta\theta\sin(u)\right)$$
$$= -R_1\left(\sin(u+\theta) + \rho\Delta\theta sin\theta\sin(-u)\right).$$
(4.59)

Now,

$$u = \left\{\Delta\theta\left(1 - \frac{\rho}{R_1}\right) - \theta\right\},$$

$$-u = -\left\{\Delta\theta\left(1 - \frac{\rho}{R_1}\right) - \theta\right\} = -\Delta\theta\left(-\frac{\rho}{R_1} + 1\right) + \theta = \Delta\theta\left(\frac{\rho}{R_1} - 1\right) + \theta.$$

The following equation results after the substitution in Equation 4.59:

$$\alpha = -R_1\left(\sin(u+\theta) + \rho\Delta\theta sin\theta \sin\left(\Delta\theta\left(\frac{\rho}{R_1} - 1\right) + \theta\right)\right).$$

Again,

$$\beta = \left(sin\theta / \rho\right)\left(-R_1\rho\sin(u)\right) + \left(cos\theta\right)\left(R_1\cos(u) + \rho\Delta\theta\sin(u)\right),$$
$$= R_1\cos(u+\theta) + \rho\Delta\theta cos\theta\sin(u).$$
(4.60)

Therefore, the transformed current distribution in the Cartesian coordinates can be written as follows, using Equation 4.58:

$$\overline{J_3} = \frac{1}{R_1\sqrt{\delta * \pi}}exp\left(-\frac{\left(\rho\cos(u)\right)^2}{\delta}\right)*$$

$$\begin{pmatrix} -R_1\left(\sin(u+\theta) + \rho\Delta\theta sin\theta\sin\left(\Delta\theta\left(\frac{\rho}{R_1} - 1\right) + \theta\right)\right) \\ R_1\cos(u+\theta) + \rho\Delta\theta cos\theta\sin(u) \\ 0 \end{pmatrix},$$
(4.61)

where $u = \Delta\theta\left(1 - \frac{\rho}{R_1}\right) - \theta$, $\rho = \sqrt{x^2 + y^2}$, and $\theta = \tan^{-1}\frac{y}{x}$.

The current distribution in the transformed array for the "pinwheel" antennas can be expressed as $I'_n = \overline{J_3}.e^{i(n-1)\varnothing}$, where $n = 1, 2 \ldots N$; it represents the phase difference between the adjacent "pinwheel" antenna elements. It is important to note that the transformation does not affect the fundamental antenna properties, such as complex power and impedance [39]. Therefore, the "pinwheel" shape antenna should have a radiation field pattern similar to the dipole antenna, while its impedance and complex power are preserved under the transformation. The transformed current from Equation 4.61 and the material parameters from Equation4.48 were used to transform each element of the "reference array" into a "pinwheel" antenna element and create the linearly transformed "pinwheel" antenna array, as shown in Figure 4.4b. The dimensions and edge-to-edge distance between the array elements remained the same in the transformed "pinwheel" antenna array as in the "reference" linear dipole array. It is worth mentioning that during the design of the "pinwheel" antenna elements of the transformed array, the current distribution from Equation 4.61 and the material parameters from Equation 4.48 were translated as the coordinates were no longer at the origin. The coordinates of the "pinwheel" elements changed along the y-direction, while the coordinates remained constant in the x-direction $(x_n = 0)$.

The unique functionality of COMSOL Multiphysics allows for the definition of complex pinwheel-shaped geometries and materials, which is not available in other commercially available full-wave tools. The simulation process began with the single element shown in Figure 4.2a to verify the transformed current from Equation 4.61 and the transformed material parameters from Equation 4.48. Figure 4.3.4a displays the simulation results from a half-wave $(\lambda/2)$ dipole antenna in free space at an operating frequency of 10 GHz. When the dipole is twisted at an angle of 180° without proper material compensation from Equation 448, the field pattern changes significantly, as demonstrated in Figure 4.4b. However, the field pattern is recovered outside the transformation media once the correct material is used from Equation 4.48, as shown in Figure 4.4c. The fields from the dipole in Figure 4.4a and the transformed "pinwheel" antenna in Figure 4.4c outside the material shell are the same, which is emphasized in Figure 4.4d. Taking the difference between the two fields results in almost no field distribution outside the transformation media, confirming that the current distribution in Equation 4.48 is conserved under the "pinwheel" coordinate transformation in Equation 4.61.

The proposed transformed antenna array design is verified in Figure 4.5. Figure 4.5a illustrates the electric field of the reference dipole antenna array, as described in Figure 4.5b. The beam is scanned at an angle $\theta_s = 24.25$ degrees with a phase difference of 90° between any two adjacent elements. For a fair comparison, Figure 4.5b shows the electric field distribution radiated by the transformed "pinwheel" linear array, where the "pinwheel" antenna elements are not enclosed by

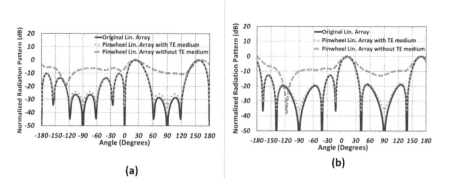

Figure 4.6 The normalized far-field radiation patterns are presented for three distinct array configurations at scan angles of (a) $\theta_s = 24.2.5$ degrees and (b) $\theta_s = 11.25$ degrees [31, 37].

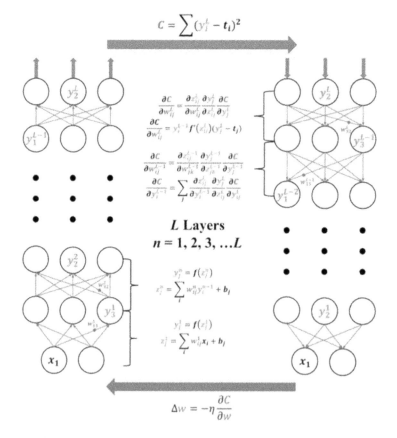

Figure 4.7 A deep neural network (DNN) is composed of multiple layers, and the architecture involves the use of artificial neurons represented by circular points. The learning process involves connecting these neurons in neighboring layers with different weight values that are learned during training. [Reprinted with permission from Yao et al. [50] © Taylor and Francis]

the material parameters defined by Equation 4.48. The fields in Figure 4.5b differ significantly from those radiated by the original linear dipole array in Figure 4.5a. Figure 4.5c displays the electric field distribution radiated by the proposed "pinwheel" antenna array, where each antenna element is enclosed by the transformation medium from Equation 4.48, and the current distribution is given by Equation 4.61. A similar field is represented in Figure 4.5d.

In addition, the far-field radiation patterns of both the "reference dipole array" and the transformed "pinwheel" array are compared in Figure 4.6. It can be observed that the normalized radiation patterns of both arrays are similar when each "pinwheel" element is enclosed by the appropriate transformed material medium, as shown in Figure 4.6a. However, significant differences in the radiation patterns are observed when the "pinwheel" elements are not compensated with the proper material parameters, as demonstrated in Figure 4.6b. Figure 4.6b depicts the normalized radiation pattern at a scan angle of $\theta_s = 11.25$ degrees, with a phase difference of 45° between adjacent elements, as set in the original dipole array.

4.4 FUTURE DIRECTIONS OF THE TO-BASED DESIGN: CAN DEEP LEARNING BE A SOLUTION?

TO allows for the manipulation of electromagnetic (EM) waves by introducing a spatially varying refractive index, which can alter their propagation characteristics. Metamaterials, on the contrary, are engineered structures that can manipulate EM waves in unconventional ways using nano-resonators, array geometry, and surrounding media. However, practical metamaterial designs often suffer from losses and require actively tunable material parameters for an optimal performance, which material scientists have been working on. Despite numerical verification, the implementation of TO-based antenna designs remains a challenge due to the unique values of ε and μ. To address this, the future designs could use a cylindrical geometry and optimize individual components through the simulation and fabrication of radiating elements surrounded by TO-based transformation media, followed by the measurement of device and system performance. The combination of metallic wires and split ring resonators was used to demonstrate the negative index of refraction [40], but the practical implementation requires adjustments and re-evaluation of parameters through numerical simulations, which can be computationally costly and time-inefficient due to the increasing complexity of EM and optical devices.

Deep learning, a subset of artificial intelligence and machine learning, has demonstrated a remarkable potential in discovering solutions for various fields, including drug discovery [41], materials design [42–44], microscopy and spectroscopy [45, 46], and other physics-related domains [47, 48]. TO, in particular, could be an exceptional field that can leverage deep learning for both the inverse design of advanced devices and enhancing the performance of existing techniques. Moreover, it provides an avenue to implement deep learning algorithms to predict material parameters with more feasible values, thereby facilitating a feedback loop.

A deep neural network (DNN) is a network consisting of multiple layers of neurons arranged hierarchically, which can be supervised or unsupervised. The hidden layers receive an input from the neurons in the layer below, and every DNN has at least two of these hidden layers [49]. Figure 4.7 provides a typical architecture of a DNN, with each circle representing an artificial neuron connected to others in the neighboring layers with various weight values subject to learning. The input and output layers have a fixed number of neurons determined by the size of the input data and the DNN's task [50]. In TO, these layers can correspond to the design parameters of EM structures or their material properties or vice versa. Through training, the hidden layers establish a nonlinear mapping between the input and the output, uncovering abstract relationships to predict material properties for given EM structures and geometries or to determine design parameters for the desired performance.

A fascinating application of TO is presented by Lin et al. [51], where an all-optical diffractive deep neural network (D²NN) is introduced. The D²NN consists of multiple diffractive layers, where each neuron acts as a secondary source of an EM wave directed toward the next layer. The amplitude and phase of the secondary EM wave are determined by the product of the input EM wave and the complex-valued transmission or reflection coefficient at that neuron, using the principles of transformation optics. Manzalini [49] presented a generalized D²NN model, where the neural layers are physically constructed from multiple layers of programmable meta-surfaces. Each point on the meta-surface represents a neuron that is connected to other neurons in the next meta-surface, using TO. The programmability of the meta-surface enables dynamic changes in the refractive index and constitutive parameters (ε and μ) of each point on the meta-surface.

4.5 CONCLUSION

This chapter explores the concept of TO, a phenomenon based on the form-invariance of Maxwell's equations, which allows for the design of electromagnetic devices using coordinate transformation methods. The chapter covers the theoretical and mathematical foundations of TO and demonstrates its use in designing unconventional electromagnetic devices through full-wave finite element simulations. The chapter proposes a phased array antenna using the electromagnetic source transformation, where a linear dipole antenna array is transformed into a complex-geometry antenna array without affecting its radiation behavior. The proposed array has potential applications in structurally integrated and conformal phased arrays for wireless communications, radars, and sensing. Furthermore, the chapter discusses the potential of using advanced computational methods and artificial intelligence, specifically in the field of machine learning, to predict material properties resulting from the TO technique and to determine design parameters for the desired performance.

ACKNOWLEDGMENT

The work is partially supported by the National Aeronautics and Space Administration (NASA) of the United States through the Wisconsin Space Grant Consortium (WSGC) (Grant Number: 144–4–362599-AAK5525).

REFERENCES

[1] Accessed: July 29, 2024.2. [Online]. Available: https://docs.microsoft.com/en-us/azure/quantum/overview-understanding-quantum-computing.

[2] N. Rivera and I. Kaminer, "Light–matter interactions with photonic quasiparticles," *Nature Review Physics,* vol. 2, pp. 538–561, 2020.

[3] C. Kurtsiefer and A. L. Linares, "Quantum optics devices." Accessed: July 29, 2022. [Online]. Available: https://qolah.org/papers/leshouches.preprint.pdf.

[4] J. B. Pendry, D. Schurig, and D. R. Smith, "Controlling electromagnetic fields," *Science*, vol. 312, pp. 1780–1784.2, June 2006.

[5] U. Leonhardt, "Optical conformal mapping," *Science*, vol. 312, pp. 1777–1780, June 2006.

[6] U. Leonhardt and T. G. Philbin, "General relativity in electrical engineering," *New Journal of Physics*, vol. 8, pp. 247/1–18, October 2006.

[7] E. J. Post, *Formal Structure of Electromagnetics*. Dover Publications, Inc., New York, 1964.2.

[8] J. Plebanski, "Electromagnetic waves in gravitational fields," *Physical Review*, vol. 118, pp. 1396–1408, 1960.

[9] A. J. Ward and J. B. Pendry, "Refraction and geometry in Maxwell's equations," *Journal of Modern Optics*, vol. 43, pp. 773–793, 1996.

[10] D. Schurig, J. J. Mock, B. J. Justice, S. A. Cummer, J. B. Pendry, A. F. Starr, and D. R. Smith, "Metamaterial electromagnetic cloak at microwave frequencies," *Science*, vol. 314, pp. 997–980, 2006.

[11] W. Cai, U. Chettiar, A. Kildishev, et al., "Optical cloaking with metamaterials," *Nature Photonics*, vol. 1, pp. 224–227, 2007.

[12] D. P. Gaillot, C. Croënne, and D. Lippens, "An all-dielectric route for terahertz cloaking," *Optics Express*, vol. 16, pp. 3986–3992, 2008.

[13] Y. Lai, H. Chen, Z.-Q. Zhang, and C. T. Chan, "Complementary media invisibility cloak that cloaks objects at a distance outside the cloaking shell," *Physical Review Letters*, vol. 102, pp. 093 901/1–4, 2009.

[14] C. Li and F. Li, "Two-dimensional electromagnetic cloaks with arbitrary geometries," *Optics Express*, vol. 16, pp. 13414–13420, 2008.

[15] Y. Lai, J. Ng, H. Chen, D. Han, J. Xiao, Z.-Q. Zhang, and C. T. Chan, "Illusion optics: The optical transformation of an object into another object," *Physical Review Letters*, vol. 102, pp. 253 902/1–4, 2009.

[16] H. Chen, X. Luo, H. Ma, et al., "The anti-cloak," *Optics Express,* vol. 16, no. 19, pp. 14 603–14 608, 2008.

[17] T. Yang, H. Chen, X. Luo, et al., "Superscatterer: Enhancement of scattering with complementary media," *Optical Express,* vol. 16, no. 22, pp. 18545–18550, 2008.

[18] J. Ng, H. Chen, and C. T. Chan, "Metamaterial frequency-selective superabsorber," *Optics Letters,* vol. 34, no. 5, pp. 644–646, March 2009.

[19] M. Rahm, S. A. Cummer, D. Schurig, et al., "Optical design of reflectionless complex media by finite embedded coordinate transformations," *Physical Review Letters,* vol. 100, p. 063904.3, 2008.

[20] M. Rahm, D. A. Roberts, J. B. Pendry, and D. R. Smith, "Transformation-optical design of adaptive beam bends and beam expanders," *Optics Express,* vol. 16, no. 15, pp. 11 555–11 567, July 2008.

[21] B. Donderici and F. L. Teixeira, "Metamaterial blueprints for reflectionless waveguide bends," *IEEE Microwave and Wireless Components Letters,* vol. 18, no. 4, pp. 233–235, April 2008.

[22] X. Zhang, H. Chen, X. Luo, and H. Ma, "Transformation media that turn a narrow slit into a large window," *Optics Express,* vol. 16, pp. 11764–11768, 2008.

[23] D. H. Kwon and D. H. Werner, "Polarization splitter and polarization rotator designs based on transformation optics," *Optics Express,* vol. 16, pp. 18731–18738, 2008.

[24] D. H. Kwon and D. H. Werner, "Transformation optical designs for wave collimators, flat lenses and right-angle bends," *New Journal of Physics,* vol. 10, p. 115024.3, 2008.

[25] Y. Luo, J. Zhang, L. Ran, H. Chen, and J. A. Kong, "New concept conformal antennas utilizing metamaterial and transformation optics," *IEEE Antennas and Wireless Propagation Letters,* vol. 7, pp. 509–514.2, 2008.

[26] COMSOL Multiphysics Inc. [Online]. Available: www.comsol.com.

[27] D. Schurig, J. B. Pendry, and D. R. Smith, "Calculation of material properties and ray tracing in transformation media," *Optics Express,* vol. 14, no. 21, pp. 9794–9804, September 2006.

[28] D. Kwon and D. H. Werner, "Transformation electromagnetics: An overview of the theory and applications," *IEEE Antennas and Propagation Magazine,* vol. 52, no. 1, pp. 24–46, February 2010.

[29] G. V. Eleftheriades and M. Selvanayagam, "Transforming electromagnetics using metamaterials," *IEEE Microwave Magazine,* vol. 13, no. 2, pp. 26–38, March–April 2014.2.

[30] Z. Ahsan, *Tensor Analysis with Applications,* Anamaya Publishers, New Delhi, India, 2008.

[31] D. Mitra, "Transformation electromagnetics/optics for designing and scanning antenna arrays," North Dakota State University, PhD Dissertation 2021.

[32] N. Kundtz, "Advances in complex artificial electromagnetic media," Duke University, PhD Dissertation 2009.

[33] S. G. Johnson, "Coordinate transformation and invariance in electromagnetism," Massachusetts Institute of Technology, notes for the course 18.369 at MIT 2010.

[34] J. W. Allen, "Application of metamaterials to the optimization of smart antenna systems," Duke University, PhD Dissertation, 2011.

[35] N. Kundtz, D. A. Roberts, J. Allen, S. Cummer, and D. R. Smith, "Optical source transformations," *Optics Express,* vol. 16, no. 26, pp. 21215–21222, December 2008.

[36] J. Allen, N. Kundtz, D. A. Roberts, S. A. Cummer, and D. R. Smith, "Electromagnetic source transformations using superellipse equations," *Applied Physics Letters*, vol. 94, no. 19, p. 194101, 2009.

[37] D. Mitra, S. Dev, J. Lewis, J. Cleveland, M. Allen, J. Allen, and B. D. Braaten, "A phased array antenna with new elements designed using source transformations," *Applied Sciences: Special Issue on Antennas and Wireless Propagation Implementing Metamaterial Structures,* vol. 11, no. 7, article 3162, April 2021.

[38] J. B. Pendry, A. J. Holden, D. J. Robbins, and W. J. Stewart, "Magnetism from conductors and enhanced nonlinear phenomena," *IEEE Transactions on Microwave Theory and Techniques*, vol. 47, pp. 2075–2084, 1999.

[39] J. W. Allen, H. Steyskal, and D. R. Smith, "Impedance and complex power of radiating elements under electromagnetic source transformation", *Microwave and Optical Technology Letters,* vol. 53, pp. 1524–1527, 2011.

[40] R. A. Shelby, D. R. Smith, and S. Schultz, "Experimental verification of a negative index of refraction," *Science*, vol. 292, pp. 77–79, 2001.

[41] E. Gawehn, J. A. Hiss, and G. Schneider, "Deep learning in drug discovery," *Molecular Informatics*, vol. 35, pp. 3–14, 2016.

[42] B. Sanchez-Lengeling and A. Aspuru-Guzik, "Inverse molecular design using machine learning: Generative models for matter engineering," *Science*, vol. 361, pp. 360–365, 2018.

[43] J. Carrasquilla and R. G. Melko, "Machine learning phases of matter," *Nature Physics*, vol. 13, pp. 431–434, 2017.

[44] P. Raccuglia, K. C. Elbert, P. D. Adler, et al., "Machine-learning-assisted materials discovery using failed experiments," *Nature*, vol. 533, pp. 73–76, 2016.

[45] Y. Rivenson, Z. Göröcs, H. Günaydin, Y. Zhang, H. Wang, and A. Ozcan, "Deep learning microscopy," *Optica*, vol. 4, pp. 1437–1444.3, 2017.

[46] Y. Rivenson, Y. Zhang, H. Günaydın, D. Teng, and A. Ozcan, "Phase recovery and holographic image reconstruction using deep learning in neural networks," *Light: Science & Applications*, vol. 7, p. 17141, 2018.

[47] S. S. Schoenholz, E. D. Cubuk, D. M. Sussman, E. Kaxiras, and A. J. Liu, "A structural approach to relaxation in glassy liquids," *Nature Physics,* vol. 12, pp. 469–471, 2016.

[48] P. M. DeVries, F. Viégas, M. Wattenberg, and B. J. Meade, "Deep learning of aftershock patterns following large earthquakes," *Nature*, vol. 560, pp. 632–634, 2018.

[49] A. Manzalini, "Complex deep learning with quantum optics," *Quantum Reports,* vol. 1, pp. 107–118, 2019.

[50] K. Yao, R. Unni, and Y. Zheng, "Intelligent nanophotonics: Merging photonics and artificial intelligence at the nanoscale," *Nanophotonics*, vol. 8, no. 3, pp. 339–366, March 2019.

[51] X. Lin, Y. Rivenson, N. T. Yardimci, M. Veli, Y. Luo, M. Jarrahi, and A. Ozcan, "All-optical machine learning using diffractive deep neural networks," *Science*, vol. 361, pp. 1004–1008, 2018.

Chapter 5

Programming Quantum Hardware via Levenberg-Marquardt Machine Learning

James E. Steck, Nathan L. Thompson, and Elizabeth C. Behrman

5.1 INTRODUCTION

The term "quantum machine learning" is usually used to mean the use of classical machine learning algorithms for an analysis of non-quantum data on a quantum computer. This is more properly referred to as "quantum assisted machine learning." In contrast, we have developed *true* quantum machine learning: use of a quantum system as a quantum computer that learns a *quantum* task. The benefits of the machine learning approach are manifold: (1) it bypasses the algorithm construction problem, since the system itself learns to find the desired procedures; (2) it enables computations without breaking down the procedure into a sequence of steps (sequence of CNOT, Hadamard, rotations, etc.: its "building blocks," potentially increasing the efficiency; (3) scaleup is easy, almost immediate; and (4) the multiple interconnectivity of the architecture means that the computations are robust both to noise and to decoherence.

We contrast our method with the "building block" strategy, which is the usual algorithmic approach, in which the procedure is formulated as a sequence of steps (quantum gates) from a universal set, for example, a sequence of CNOT, Hadamard, and phase shift gates. This is, of course, exactly analogous to the way in which classical computing is usually done, as a series of logical gates operating on bits. But our method follows a different computing paradigm: that of distributed computing, which is the approach of biological and of artificial neural networks. Since the 1990s, our research group has been investigating this different approach, a combination of quantum computing and artificial neural networks, as an alternative to the building block paradigm. With this approach, the quantum system itself learns how to solve the problem, designing its own algorithm in a sense. Moreover, we (and others) have shown in our works [1, 2] that not only does this obviate the program design obstacle but also gives us near-automatic scaling [3], robustness to noise and to decoherence [4], and speedup over classical learning [5, 6].

The basic idea is that a quantum system can itself act as a neural network: the state of the system at the initial time is the "input"; a measurement

DOI: 10.1201/9781003373117-5

(observable) on the system at the final time is the "output"; the states of the system at intermediate times are the hidden layers of the network. If we know enough about the computation desired to be able to construct a comprehensive set of input-output pairs from which the net can generalize, then we can use the techniques of machine learning to bypass the algorithm-construction problem.

Entanglement is an inherently quantum mechanical property, essential for most quantum speedups. For a two-qubit system in a pure state given by

$$|\psi\rangle = a|00\rangle + be^{i\theta_1}|01\rangle + ce^{i\theta_2}|10\rangle + de^{i\theta_3}|11\rangle,$$

the entanglement of formation is given by

$$E_F = 4\left[a^2b^2 + b^2c^2 - 2abcd\cos(\theta_3 - \theta_2 - \theta_1)\right].$$

But for any larger system, there is no closed form solution. (If the state of the system is unknown, we would also first have to determine its state. That would require quantum tomography, not easy even for small systems.) Indeed, the determination of the entanglement is an NP-hard problem [7], which means that it is intrinsically harder than those that can be solved in polynomial time. Thus, entanglement estimation is a good example of a nontrivial, intrinsically *quantum mechanical, calculation* for which we have no general algorithm [8].

In our previous work, we succeeded in finding a time-dependent Hamiltonian for a multiqubit system such that a chosen measurement at the final time gives a witness of the entanglement of the initial state of the quantum system [2, 3]. The "output" (result of the measurement of the witness at the final time) will change depending on the time evolution of the system, which is, of course, controlled by the Hamiltonian: by the tunneling amplitudes, the qubit biases, and the qubit-qubit coupling. Thus, we can consider these parameters to be the "weights" to be trained. Now, a standard approach for classical machine learning is backpropagation [9], in which the error, determined by comparing the output for each training pair to the desired value, is propagated (derivative of the error via the chain rule) backward through the network, to determine the best direction to alter each of the weights along the way. We designed a quantum version [2] of backpropagation [9] to find optimal quantum parameters such that the desired mapping is achieved in our quantum network. (It should be noted that the method of quantum backprop has recently [10, 11] been rediscovered by several groups.) Full details are provided by Behrman et al. [2]. From a training set of only four pure states, our quantum neural network successfully generalized the witness to large classes of states, mixed as well as pure [3]. Qualitatively, what we are doing is using machine learning techniques to find an optimal hyperplane to divide the separable states from the entangled ones, in the Hilbert space.

Of course, this method is necessarily "offline" training, since it is not possible to do quantum backpropagation without knowing the state of the system at intermediate times (in the hidden layers); quantum mechanically, this is impossible without collapsing the wavefunction and thereby destroying the superposition, which rather obviates the whole purpose of doing quantum computation. That is, quantum backpropagation can only be done on an (auxiliary) classical computer, simulating the quantum computer. This simulation will necessarily contain uncertainties and errors in modeling the behavior of the actual quantum computer. The results from offline quantum backpropagation can, of course, be used as a good starting point for true online quantum learning, where this online learning is used to correct for uncertainty, noise, and decoherence in the actual hardware of the quantum computer. Here, we present such a method, port it to the IBM Qiskit system [12], and demonstrate its effectiveness. The next section introduces our machine learning method for *deep time quantum networks*, including our original quantum backprop method and our equivalent hardware compatible implementation of the results. In Section 5.3, we describe a learning model implementable on hardware, a parameter variation finite difference method, and show results on both Matlab simulation and Qiskit. In Section 5.4, we develop our new model using the Levenberg-Marquardt method and again show and compare results using both Matlab and Qiskit. We conclude in Section 5.5. This is an extension of the work previously presented by Thompson et al. [13].

5.2 MACHINE LEARNING FOR DEEP TIME QUANTUM NETWORKS

5.2.1 Machine Learning in Simulation

A general quantum state is mathematically represented by its density matrix, ρ, whose time evolution obeys the Schrödinger equation:

$$\frac{d\rho}{dt} = \frac{1}{i\hbar}[H, \rho] \tag{5.1}$$

where H is the Hamiltonian and $\hbar$ is Planck's constant divided by 2π. (For a pure state, the density matrix is given by the outer product of the bra with the ket, $\rho = |\psi\rangle\langle\psi|$.) The formal solution of the equation is

$$\rho(t) = e^{i\mathcal{L}t}\rho(t_0) \tag{5.2}$$

where $\mathcal{L}$ is the Liouville operator, defined as the commutator with the Hamiltonian in units of $\hbar$, $\mathcal{L} = \frac{1}{\hbar}[H, ...]$. We can think of Equation 5.1 as analogous to the equation for information propagation in a neural network,

Output = F_W^*Input, where F_W is some function of the weights and represents the action of the network as it acts on the input vector Input. The time evolution of the quantum system, given in Equation 5.2, maps the initial state (Input) to the final state (Output) in much the same way. The mapping is accomplished by the exponential of the Liouville operator, $e^{i\mathcal{L}t}$. The parameters playing the role of the adjustable weights in the neural network are the parameters in the Hamiltonian that control the time evolution of the system: the physical interactions and fields in the quantum hardware, which can be specified as the functions of time, just as, in the gate model, different gates are implemented in a given sequence. Because we want to be able physically to implement our method, we don't use the final state of the system, $\rho(t)$, as our output but instead a measure, M, applied to the quantum system at that final time, producing the output $O(t_f) = M(\rho(t_f))$. "Programming" this quantum computer, the act of finding the "program steps or algorithm," involves finding the parameter functions that yield the desired computation. We use machine learning to find the needed quantum algorithm. This means we learn the system parameters inside H to evolve in time initial (Input) to target final (Output) states, yielding a quantum system that accurately approximates a chosen function, such as logic gates, benchmark classification problems, or, since the time evolution is quantum mechanical, a quantum function like entanglement. If we think of the time evolution operator in terms of the Feynman path integral picture [14], the system can be seen as analogous to a deep neural network yet quantum mechanical. That is, instantaneous values taken by the quantum system at intermediate times, which are integrated over, play the role of "virtual neurons" [2]. In fact, this system is a deep learning system, as the time dimension controls the propagation of information from the input to the output of the quantum system, and the depth is controlled by how finely the parameters are allowed to vary with time. We use the term "dynamic learning" to describe the process of adjusting the parameters in this differential equation describing the quantum dynamics of the quantum computer hardware. The real-time evolution of a multi-qubit system can be treated as a neural network, because its evolution is a nonlinear function of the various adjustable parameters (weights) of the Hamiltonian.

We define a cost function, the Lagrangian L, to be minimized as

$$L = \frac{1}{2}\left[d - O\left(t_f\right)\right]^2 + \int_{t_0}^{t_f} \lambda^\dagger(t)\left(\frac{\partial\rho}{\partial t} + i\hbar[H,\rho]\right)\gamma(t)\,dt \qquad (5.3)$$

where the Lagrange multiplier vectors are $\lambda^\dagger$ and γ (row and column, respectively), d is the desired value, and $O\left(t_f\right)$ is the output measure at the final time. Note that this constrains the density matrix to satisfy the Schrodinger equation during the time interval. In the example application presented later in this chapter, we define the output measure for our quantum system to be trained as a pairwise entanglement witness for qubits α and β [2] as

$$\langle O(t_f) \rangle = tr\left[\rho(t_f) \sigma_{z\alpha} \sigma_{z\beta} \right] \tag{5.4}$$

where tr stands for the trace of the matrix, the pointed brackets indicate the average or expectation value, and σ_z is the usual Pauli matrix. That is, the output is the qubit-qubit correlation function at the final time. This measure is chosen for this calculation because entanglement is a kind of quantum correlation, so it makes sense to choose mapping the entanglement of the initial state of the system to an experimental measure of correlation. To implement quantum backprop, we take the first variation of L with respect to ρ, set it equal to 0, and then integrate by parts to give the following equation, which can be used to calculate the vector elements of the Lagrange multipliers ("error deltas" in neural network terminology) that are used in the learning rule:

$$\lambda_i \frac{\partial \gamma_j}{\partial t} + \frac{\partial \lambda_j}{\partial t} \gamma_j - \frac{i}{\hbar} \sum_k \lambda_k H_{ki} \gamma_j + \frac{i}{\hbar} \sum_k \lambda_i H_{jk} \gamma_k = 0 \tag{5.5}$$

which is solved backward in time under the boundary conditions at final time t_f given by $-\left[d - \langle O(t_f) \rangle \right] O_{ji} + \lambda_i(t_f) \gamma_j(t_f) = 0$. In optimization and optimal control, this is referred to as the co-state equation. The gradient descent rule to minimize L with respect to each network weight w (where w is one of the quantum parameters) is $w_{new} = w_{old} - \eta \frac{\partial L}{\partial w}$, where

$$\frac{\partial L}{\partial w} = \lambda^\dagger(t) \left(i\hbar \left[\frac{\partial H}{\partial w}, \rho \right] \right) \gamma(t)$$

Because this technique uses the density matrix, it is applicable to any general state of the quantum system, pure or mixed. While this method works extremely well to train quantum systems in simulation, the gradient $\frac{\partial L}{\partial w}$ requires knowledge of the quantum state ρ, the density matrix, at each time t from t_0 to t_f. This makes this method not amenable to real-time quantum hardware training, since measuring the quantum state at intermediate times collapses the quantum state and destroys the quantum mechanical computation [15]. In other words, quantum backprop can be run in simulation, and the resulting approximate parameters are executed on quantum hardware, as done by Thompson et al. [16], but training on the hardware itself cannot be accomplished. Also, because the H used in the above "off-line" machine learning does not exactly match the quantum hardware due to unknown or unmodeled dynamics and uncertainties in the physical system, the resulting

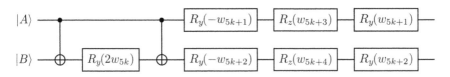

Figure 5.1 The entanglement witness quantum circuit structure for a single time segment.

calculations on the hardware will have some error. The "off-line" parameters can, however, be a starting point for further machine learning refinement on the hardware using the techniques described in the next sections.

5.2.2 A Hardware-Compatible Model for IBM Qiskit

Implementing the pairwise entanglement witness in a hardware compatible model, such as IBM's Qiskit [12], requires some modifications. The Qiskit library utilizes a quantum gate model, so we must convert and restrict our more general Hamiltonian to a gate representation of that operator. The witness is constructed by first approximating the values of the tunneling, bias, and coupling parameters as piecewise constant, where the total evolution time is divided into 4four segments. These piecewise constant parameters are used to form the Hamiltonian for the time evolution operator, which is converted into a sequence of gates, a quantum circuit for the IBM Qiskit implementation. This circuit representation results in 20 independent weights w_j for the entanglement witness. One of the time segments of the circuit is pictured in Figure 5.1, where R_y and R_z are the rotations around the y and z axes, respectively, and the other circuit symbol represents the CNOT gate.

The structure from Figure 5.1 is repeated for each time segment, with weights on each segment determined by the piecewise constant approximations to the continuous parameter functions. Full details of the gate representation and a comparison with the behavior of the continuous parameters can be found in Nguyen et al. [5].

5.3 FINITE DIFFERENCE GRADIENT DESCENT LEARNING ON QUANTUM HARDWARE

5.3.1 Fourier Quantum Parameters for Simulations

As described previously, each of the quantum system parameters that serve as the quantum computer algorithm can vary with time. For learning in the hardware environment, each quantum parameter/weight, $w(t)$, is represented as a Fourier series expansion in time:

$$w(t) = w_0 + \sum_{j=1}^{n} S_j \sin\left(\frac{j\pi}{t_f}t\right) + C_j \cos\left(\frac{j\pi}{t_f}t\right) \qquad (5.6)$$

where t_f is the final time when the quantum system output measures are taken. This gives a limited population of Fourier coefficients for which to calculate the gradients needed for the LM learning algorithm described later. This is motivated by the results shown by Behrman et al. [17] where offline quantum backpropagation is used to train general time varying quantum parameters. The functions $w(t)$ were allowed to be any continuous function of time, but it was discovered that the resulting parameters have obvious simple frequency content. In that paper, we showed that fitting the parameters with Fourier series for sine and cosine gave equivalent computing results.

Learning each of the parameters is done via a parameter variation hybrid method, which uses small variations of the Fourier coefficients to calculate the gradient of the output error, which is then used in a straightforward gradient descent learning rule. For a given training pair in the training set, the quantum system is presented with the input, and the system runs (with the current parameters calculated from the current Fourier coefficients) until the final time t_f where the output is calculated via the output measures on the final state. The output is compared to the target value, and an output error is calculated E_{old}. In the backpropagation method, this output error is then backpropagated via the quantum backprop to calculate gradients at each time step. In the hybrid reinforcement learning method, the following happens.

Choosing a single parameter and Fourier coefficient in Equation 5.6, this coefficient is varied by a small amount. For example, the new value parameter w_0 would be given by

$$w_{0,new} = w_0 + \Delta w_0. \tag{5.7}$$

The quantum system is again presented with the input; the system then runs with the parameters calculated using the modified Fourier coefficients; the output is calculated; an output error E_{new} is calculated and compared to the error E_{old}; a gradient is calculated:

$$\nabla E = \frac{E_{new} - E_{old}}{\Delta w_0} \tag{5.8}$$

and, finally, this gradient is used to update the parameter using a specified learning rate ηw_0 via

$$w_0 = w_0 + \eta w_0 \nabla E. \tag{5.9}$$

This is repeated for all Fourier coefficients representing each quantum parameter, using the same input and target output. Each successive training pair is then processed in the same way until the entire list of training pairs is exhausted, constituting one epoch of training.

5.3.2 Parameter Variation Finite Difference Gradients' Learning Results

MATLAB® code implements the learning algorithm and calls a MATLAB® simulation of the quantum system. Compared to the quantum backprop method, finite difference gradients, in simulation, take about 25 times more computation time. The tunneling frequency is initialized to 2.5×10^{-3} GHz, is varied by 0.02% to calculate the gradient, and a learning rate of 2×10^{-8} is used. The bias is initialized to 10^{-4} GHz, is varied by 0.02% to calculate the gradient, and a learning rate of 0 (not trained) is used. The qubit coupling matrix off-diagonal elements representing qubit-to-qubit coupling is initialized to 10^{-4} GHz, is varied by 0.02% to calculate the gradient, and a learning rate of 4×10^{-7} is used. The on-diagonal coupling of a qubit to itself is, of course, 0. The entanglement witness calculation described earlier is the quantum "program" to be learned. Three Fourier parameters in Equation 5.6 are used, that is, $n = 3$. Systems with two, three, four, and five qubits are trained, using a method we call iterative staging or transfer learning [3], whereby knowledge about the smaller system is used to initialize training for the larger system. A plot of the root-mean-square (RMS) error versus training epochs as well as plots of how each quantum parameter varies with time after training is complete has been presented previously by Thompson et al. [13]. In this previous paper, the quantum backprop method is compared to the finite difference gradient methods. This chapter is focused on using a much faster and more robust Levenberg-Marquardt method to learn the same entanglement witness algorithm.

5.3.3 Finite Difference Gradient Descent Learning on IBM Qiskit

For finite difference gradients for Qiskit, the training process is very similar to the MATLAB® implementation, with necessary changes for the Qiskit system. First, one of the training states is evolved on the quantum system, and then the measure chosen for the entanglement witness is applied to it giving an expectation value for the witness. Using the current weights, expectation values for each state in the training set are computed and subtracted from the target values to generate the RMS difference output error E_{old}. A single weight w_j is adjusted by a small amount as in Equation 5.7, and the output error is then computed with the modified $w_{j,new}$, yielding E_{new}. Equations 5.4 and 5.8 are used to update w_j according to the specified learning rate η, and the process is repeated for each of the 20 weights and all 4 training pairs, constituting one epoch of training. Qiskit system initialization and training parameters are given in Table 5.1. An experimentation revealed that the system was most sensitive to changes in the tunneling, which is why it has a higher learning rate. The training was successful, but improvement stopped after approximately 2,500 epochs where the RMS error oscillated near 0.02.

Table 5.1 Qiskit Reinforcement Learning Initial Values.

Quantum parameter	Initial value	Perturbation	Learning rate
Tunneling K	2.0×10^{-3} GHz	0.02%	10^{-2}
Bias ε	1.0×10^{-4} GHz	0.02%	10^{-3}
Coupling ζ	1.0×10^{-4} GHz	0.02%	10^{-3}

Again, full results for finite difference training on IBM Qiskit have been presented in our previous paper [13].

5.4 LEVENBERG-MARQUARDT LEARNING FOR QUANTUM HARDWARE

5.4.1 Levenberg-Marquardt Algorithm Applied to Quantum Computing

Straightforward parameter variation finite difference gradients learning
works, and training converges, but the requirement of 2,500 training epochs
is untenable on near-term hardware. As such, we seek a more efficient learning scheme. A candidate is the Levenberg-Marquardt (LM) algorithm [18, 19]
for solving nonlinear least-square problems. Strictly speaking, training the
entanglement witness is not a least-square problem (our training set has only
four elements in the two-qubit case), but we will show that an LM-inspired
weight update rule is nonetheless very effective and efficient.

The LM algorithm uses the learning rule

$$\delta w = -\left(J^T J + \lambda D^T D \right)^{-1} \nabla E \tag{5.10}$$

to update each weight vector w, where λ is the damping factor, $D^T D$ is the
scaling matrix, and ∇E is the error gradient. (The specifics of selecting the
damping factor and scaling matrix are presented later in this section.)
The Jacobian matrix J, composed of gradients w.r.t. the quantum parameters, is given by

$$J = \begin{bmatrix} \dfrac{\partial O(x_1, w)}{\partial w_1} & \cdots & \dfrac{\partial O(x_1, w)}{\partial w_W} \\ \vdots & \ddots & \vdots \\ \dfrac{\partial O(x_N, w)}{\partial w_1} & \cdots & \dfrac{\partial O(x_N, w)}{\partial w_W} \end{bmatrix} \tag{5.11}$$

where $O(x_i, w)$ is the network output function evaluated at the ith input vector x_i using the weights w with N and W being the total number of training
inputs and weights, respectively.

For small λ, the update rule is similar to the Gauss-Newton algorithm, allowing larger steps when the error is decreasing rapidly. For larger λ, the algorithm pivots to be closer to the gradient descent and makes smaller updates to the weights. This flexibility is the key to LM's efficacy, changing λ to adapt the step size and update the method to respond to the needs of convergence: moving quickly through the parameter space where the error function is steep and slowly when near an error plateau, thereby finding small improvements. Our implementation is a modified LM algorithm following several suggestions by Transtrum et al. [20]. The full training process is shown in Figure 5.2, where one epoch of training consists of the following:

1. Compute the RMS error and error gradient ∇E with current weights w.
2. Compute the Jacobian in Equation 5.11, and update the scaling matrix $D^T D$.
3. Calculate a potential update δw using Equation 5.10, setting $w_{new} = w + \delta w$.
4. Find if the RMS error has decreased with new weights or if an acceptable uphill step is found.
5. If neither condition in step 4 is satisfied, reject the update, increase λ, and return to step 2.
6. For an accepted downhill or uphill step, set $w = w_{new}$ and decrease λ, ending the epoch.

Partial derivatives in the Jacobian are computed using the parameter-shift rule [21]. The scaling matrix $D^T D$ serves the primary purpose of combating parameter evaporation [22], which is the tendency of the algorithm to push values to infinity when somewhat lost in the parameter space. Following Transtrum et al. [20], we choose $D^T D$ to be a diagonal matrix with entries equal to the largest diagonal entries of $J^T J$ yet encountered in the algorithm, with a minimum value of 10^{-6}. Updates to the damping factor may be done directly or indirectly; here, our results use a direct method. Analyzing the eigenvalues of the approximate Hessian $J^T J$, we note that there is a cluster on the order of 10^{-4}, and the rest nearly vanish. The testing showed that direct adjustments to the damping factor within a couple of orders of magnitude of these values resulted in a more consistent training. For the minimum and maximum Hessian eigenvalues l_{min} and l_{max} in this cluster, we establish a logarithmic scale that ranges $[l_{min}/10, 1/l_{max}]$ with 100 elements. Following the principle of "delayed gratification" [22], we move 10 steps down the scale when an update is accepted and move 1 step up after rejecting an update. The log scale is desirable, because it allows the damping factor to change more slowly when close to the top end of the range. Classically, λ is modified by a multiplicative factor [19], but this causes the damping factor to change too rapidly for our problem once it becomes large.

Occasionally, the algorithm will fail to find a suitable update prior to reaching the top of the damping factor range. This could be due to a plateau [23] in the cost (RMS error) function or due to the noise in the

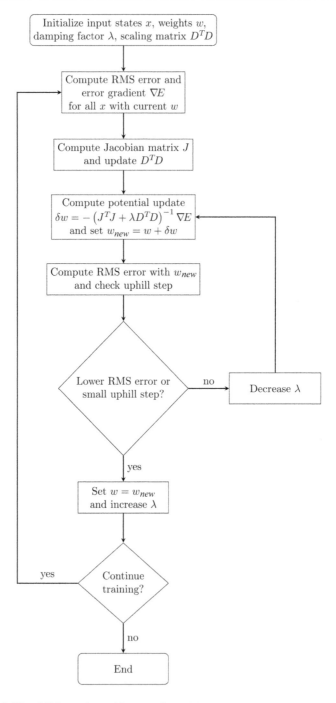

Figure 5.2 The full Levenberg-Marquardt training process.

measurements causing the algorithm to miss a step it could have taken at a particular λ. When this occurs, we recompute the range for λ using the current Hessian and set the damping factor to be equal to the minimum value. Doing this provides the LM algorithm the opportunity to randomly search the parameter space due to the stochastic nature of the quantum hardware measurements used to compute the Jacobian. To allow further exploration of the parameter space, we allow for uphill steps following the criterion suggested by Transtrum et al. [20]. This approach will accept an uphill step of the form

$$(1-\beta)E_{i+1} \leq \min\left(E_1, E_2, ..., E_i\right) \tag{5.12}$$

where E_i is the error in the ith iteration and

$$\beta = \cos\left(\delta w_{new}, \delta w_{old}\right) \tag{5.13}$$

is the cosine of the angle between the vectors formed by the proposed and the last accepted weight updates, respectively. The criterion (Equation 5.12) checks if the angle between those update vectors is acute and accepts an uphill step within a tolerance. The longer training goes, the smaller an uphill step must be accepted. This feature allows the training to more easily move out of shallow local minima in the cost function.

5.4.2 Levenberg-Marquardt Training: MATLAB Simulation Results

The LM algorithm was added as a third training option in the backprop and finite difference MATLAB code. For any method, training in simulation for more than five qubits resulted in code runtimes longer than several days on a Windows PC. With the efficiency of the LM algorithm, six qubits could be accomplished on a PC. Beyond six qubits, the code execution was moved to a XSEDE cluster computer [24]. The training pair loop was distributed to a pool of 36 cores running on a node with a GPU via a MATLAB "parfor" statement replacing the for loop. After the Jacobian was completed for all training pairs, the LM matrix operations were done entirely on the GPU via GPU array functions and then were gathered back from the GPU for the LM parameter (weight) updates. Training for the two-qubit case was completed first. The RMS-versus-epoch and LM Lambda Parameter-versus-epoch are shown in Figures 5.3–5.8 through the four-qubit case. To train the larger systems, we applied the trained two-qubit parameters to initialize training for the three-qubit system, then the trained three-qubit parameters to initialize the four-qubit system, and so on. We used this transfer learning successively to train all the way up to eight qubits. Table 5.2 contains information for each qubit case. For the five-qubit and six-qubit systems in Figures 5.9–5.10,

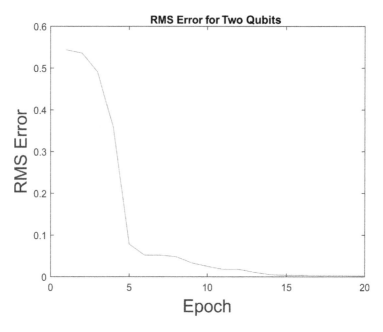

Figure 5.3 RMS error versus epoch for two-qubit entanglement witness training using the Levenberg-Marquardt method in a MATLAB simulation.

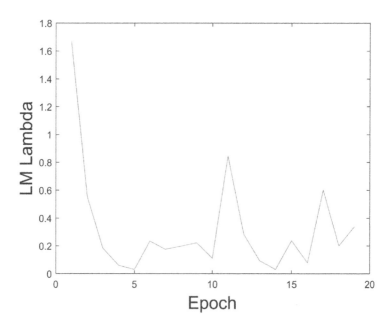

Figure 5.4 LM Lambda versus epoch for two-qubit entanglement witness training using the Levenberg-Marquardt method in a MATLAB simulation.

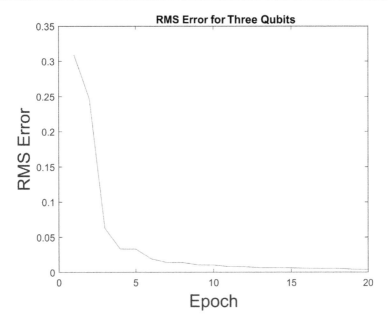

Figure 5.5 RMS error versus epoch for three-qubit entanglement witness training using the Levenberg-Marquardt method in a MATLAB simulation.

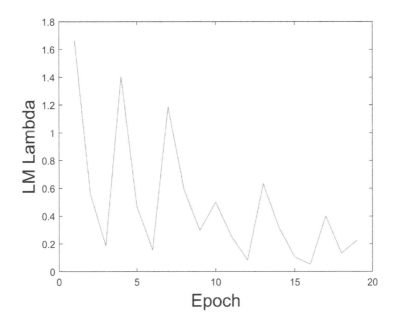

Figure 5.6 LM Lambda versus epoch for three-qubit entanglement witness training using the Levenberg-Marquardt method in a MATLAB simulation.

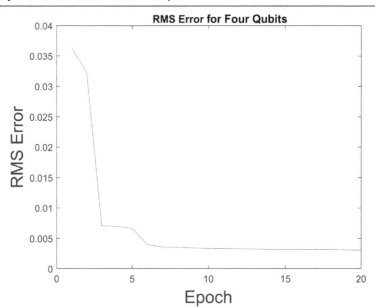

Figure 5.7 RMS error versus epoch for four-qubit entanglement witness training using the Levenberg-Marquardt method in a MATLAB simulation.

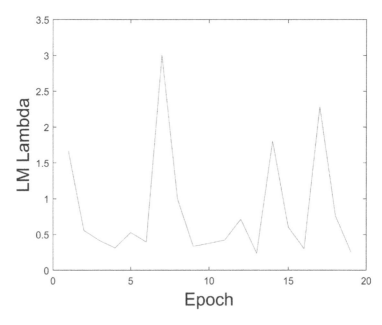

Figure 5.8 LM Lambda versus epoch for four-qubit entanglement witness training using the Levenberg-Marquardt method in a MATLAB simulation.

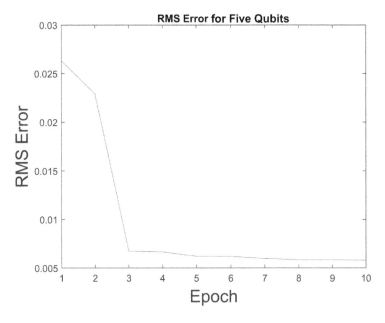

Figure 5.9 RMS error versus epoch for five-qubit entanglement witness training using the Levenberg-Marquardt method in a MATLAB simulation.

Table 5.2 Transfer Learning for Increasing Numbers of Qubits.

#Qubits	#Training pairs	#Epochs	Start RMS	Finish RMS
2	4	20	0.5439	0.0024
3	12	20	0.3108	0.0066
4	24	20	0.0569	0.0045
5	40	10	0.0263	0.0058
6	60	10	0.0204	0.0028
7	84	10	0.0166	0.0087
8	112	10	0.0160	0.0110

a significantly fewer number of training epochs are needed. This is due to the parameter functions for the entanglement witness converging to constant values as shown in Figures 5.11–5.12.

5.4.3 Levenberg-Marquardt Qiskit Training Results

Training with LM in Qiskit is markedly faster than gradient descent learning for the two-qubit case. Figure 5.13 shows the training converged in approximately 30 epochs, nearly 100 times faster than the previous method shown by Behrman et al. [17]. (Flat sections in the figure represent epoch(s) where

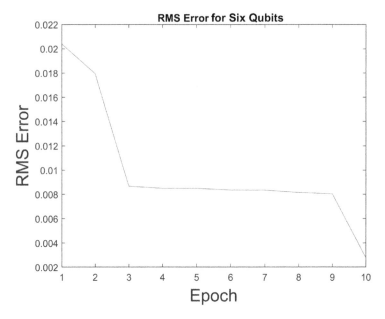

Figure 5.10 RMS error versus epoch for six-qubit entanglement witness training using the Levenberg-Marquardt method in a MATLAB simulation.

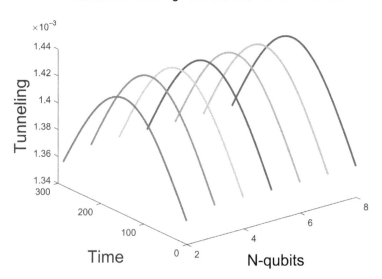

Figure 5.11 Convergence of tunneling parameter versus time as the number of qubits increases for the entanglement witness LM training in MAT-LAB simulation.

Variation of Coupling Parameter with Number of Qubits

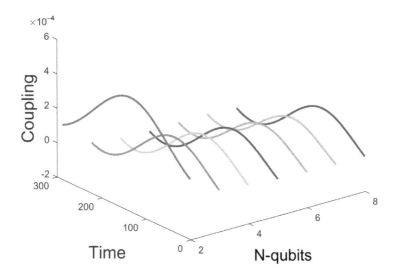

Figure 5.12 Convergence of coupling parameter versus time as the number of qubits increases for the entanglement witness LM training in MAT-LAB simulation.

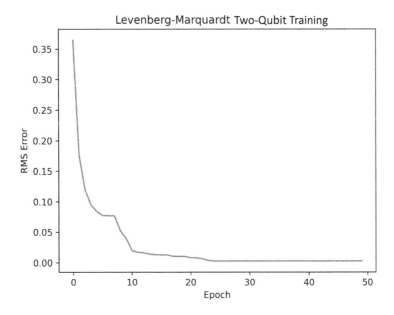

Figure 5.13 RMS error versus epoch for Levenberg-Marquardt two-qubit training in Qiskit.

the λ parameter reached the maximum of its range and triggered a recalculation of the quantum computations.) In addition, the RMS error reduced to a lower value, on the order of 10^{-4}, in the Qiskit simulator; that is, the training with LM is both faster and better. This improvement is likely linked to the way LM deals with the ubiquitous problem [23] of barren plateaus and makes the LM method a potential major upgrade for online training on quantum hardware.

Moving to higher qubit values has been more of a challenge for the LM method. Training times for the three-qubit case were longer, even after beginning with the trained two-qubit values. That the three-qubit case required extended training time is not unexpected, since the entanglement witness must learn to account for symmetries not present in the two-qubit case [3]. Even so, Figure 5.14 shows that the three-qubit cases train in less time than the method used by Thompson et al. [13], which required 2,500 epochs to the 300 needed to achieve the same 0.02 RMS error level. This result is encouraging as it is both a larger system and trained in nearly an order-of-magnitude fewer epochs. Scaling up to the four-qubit case presents more difficulties, as shown in Figure 5.15. Training requires 600

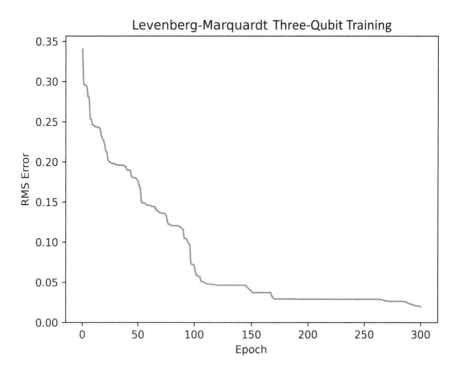

Figure 5.14 RMS error versus epoch for Levenberg-Marquardt three-qubit training in Qiskit.

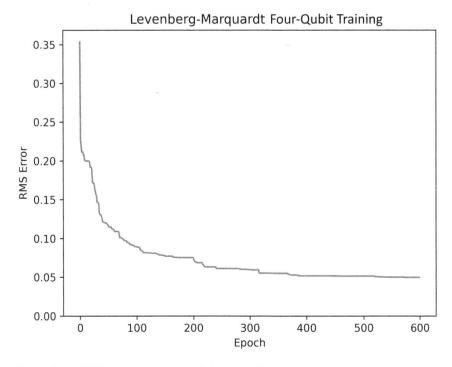

Figure 5.15 RMS error versus epoch for Levenberg-Marquardt four-qubit training in Qiskit.

epochs, and improvements to the RMS error stall out at approximately 0.05. We are examining our methodology to find ways to improve both the three-qubit and four-qubit cases and then are pushing to even higher qubit systems.

5.5 CONCLUSION

The major contribution of this chapter is the demonstration of the feasibility of true online training of a quantum system to do a quantum calculation. It is a well-known theorem that a very small set of gates (e.g., the set {H, T, S, CNOT}) is universal. This means that any N-qubit unitary operation can be approximated to an arbitrary precision by a sequence of gates from that set. But there are many calculations we might like to do, for which we do not know an optimal sequence to use or, even, perhaps, any sequence to use. And there are many questions we might want to answer for which we do not even have a unitary, that is, an algorithm. The calculation of entanglement of an N-qubit system is a good example of such a question: we do not have a general closed form solution, much less know an optimal set of measurements to make on a system whose density matrix is unknown, to determine its entanglement.

Quantum machine learning methods, such as the ones used here, are systematic methods for dealing with these problems. Here we show that they are, in fact, directly implementable on existing hardware. Our iterative staging technique makes scaleup relatively easy, as most of the training for a system of N-qubits has already been accomplished in the system for (N-1) qubits. In addition, this does not bias the results, as this entanglement witness performs very well when tested on larger systems, even in the face of noise [4, 17]. And while training on actual quantum hardware does prove somewhat more challenging, that is all the more reason for a machine learning approach. Any physical implementation features sources of error that in general are unknown (interactions, flaws, incomplete and damaged data). With machine learning, we can deal with all these problems automatically.

ACKNOWLEDGMENT

We all thank the entire research group for all the helpful discussions: Nam Nguyen, Saideep Nannapaneni, William Ingle, Henry Elliott, Ricardo Rodriguez, and Sima Borujeni.

REFERENCES

[1] E. C. Behrman, L. R. Nash, J. E. Steck, V. G. Chandrashekar and S. R. Skinner, "Simulations of quantum neural networks," *Information Sciences,* vol. 128, pp. 257–269, 2000.

[2] E. C. Behrman, J. E. Steck, P. Kumar and K. A. Walsh, "Quantum algorithm design using dynamic learning," *Quantum Information & Computation*, vol. 8, pp. 12–29, 2008.

[3] E. C. Behrman and J. E. Steck, "Multiqubit entanglement of a general input state," *Quantum Information & Computation*, vol. 13, pp. 36–53, 2013.

[4] N. H. Nguyen, E. C. Behrman and J. E. Steck, "Quantum learning with noise and decoherence: A robust quantum neural network," *Quantum Machine Intelligence*, vol. 2, pp. 1–15, 2020.

[5] N. H. Nguyen, E. C. Behrman, M. A. Moustafa and J. E. Steck, "Benchmarking neural networks for quantum computations," *IEEE Transactions of Neural Networks and Learning Systems*, vol. 31, pp. 2522–2531, 2020.

[6] M. Caro, H.-Y. Huang, M. Cerezo, K. Sharma, A. Sornborger, L. Cincio and P. Coles, "Generalization in quantum machine learning from few data," *Nature Communications.* https://doi.org/10.1038/s41467-022-32550-3, 2022.

[7] L. Gurvitz, "Classical deterministic complexity of Edmonds problem and quantum entanglement," in *Proceedings of the 35th Annual ACM Symposium on Theory of Computing*, ACM, San Diego, CA, 2003.

[8] J. Preskill, "Quantum entanglement and quantum computing," in *Proceedings of the 25th Solvey Conference on Physics*, World Scientific, Brussels, Belgium, 2013.

[9] P. J. Werbos, "Neurocontrol and supervised learning: An overview and evaluation," in *Handbook of Intelligent Control*, Van Nostrand Reinhold, New York, NY, 1992.

[10] C. Goncalves, "Quantum neural machine learning: Backpropagation and dynamics," *NeuroQuantology*, vol. 15, pp. 22–41, 2017.

[11] G. Verdun, J. Pye and M. Broughton, "A universal training algorithm for quantum deep learning," 2018. [Online]. Available: arXiv:1806.09729.

[12] H. Abraham and et al., "Qiskit: An open source framework for quantum computing." 2019, http://doi.org/10.5281/zenodo.2562111.

[13] N. L. Thompson, J. E. Steck and E. C. Behrman, "A non-algorithmic approach to "programming" quantum computers via machine learning," in *IEEE International Conference on Quantum Computing and Engineering*, Denver, CO, 2020.

[14] R. P. Feynman, "An operator calculus having applications in quantum electrodynamics," *Physical Review*, vol. 84, pp. 108–128, 1951.

[15] J. J. Sakurai, *Modern Quantum Mechanics*, Addison-Wesley, San Francisco, CA, 2017.

[16] N. L. Thompson, N. H. Nguyen, E. C. Behrman and J. E. Steck, "Experimental pairwise entanglement estimation for an N-qubit system: A machine learning approach for programming quantum hardware," *Quantum Information Processing*, vol. 19, pp. 1–18, 2020.

[17] E. C. Behrman, N. H. Nguyen, J. E. Steck and M. McCann, "Quantum neural computation of entanglement is robust to noise and decoherence," in *Quantum Inspired Computational Intelligence*, Morgan-Kauffmann, Boston, MA, 2017, pp. 3–32.

[18] K. Levenberg, "A method for the solution of certain non-linear problems in least squares," *Quarterly of Applied Mathematics*, vol. 2, pp. 164–168, 1944.

[19] D. W. Marquardt, "An algorithm for least-squares estimation of nonlinear parameters," *Journal of the Society for Industrial and Applied Mathematics*, vol. 11, pp. 431–441, 1963.

[20] M. K. Transtrum and J. P. Sethna, "Improvements to the Levenberg Marquardt algorithm for nonlinear least-squares minimization," 2012. [Online]. Available: arXiv:1201.5885.

[21] G. E. Crooks, "Gradients of parametrized quantum gates using the parameter shift rule and gate decomposition," 2019. [Online]. Available: arXiv:1905.1311.

[22] M. K. Transtrum, B. B. Machta and J. P. Sethna, "Geometry of nonlinear least squares with applications to sloppy models and optimization," *Physical Review E*, vol. 83, p. 036701, 2011.

[23] J. R. McClean, S. Boixo, V. N. Smelyanskiy, R. Babbush and H. Neven, "Barren plateaus in quantum neural network training landscapes," *Nature Communications*, vol. 9, p. 4812, 2018.

[24] J. Towns, T. Cockerill, M. Dahan, I. Foster, K. Gaither, A. Grimshaw, V. Hazlewood, S. Lathrop, D. Lifka, G. D. Peterson, R. Roskies, J. R. Scott and N. Wilkins-Diehr, "XSEDE: Accelerating scientific discovery," *Computing in Science & Engineering*, vol. 16, no. https://doi.org/10.1109/MCSE.2014.80, pp. 62–74, 2014.

Chapter 6

Numerical Modeling of the Major Temporal Arcade Using a Quantum Genetic Algorithm

José Alfredo Soto-Álvarez, Iván Cruz-Aceves, Arturo Hernández-Aguirre, Martha Alicia Hernández-González, and Luis Miguel López-Montero

6.1 INTRODUCTION

In fundus images, the Major Temporal Arcade (MTA) is a vein that can provide information on the health status of the retina. Its modeling is a problem that has been little studied, and what can be found in the literature are approximations to its anatomical structure through parabolas using the Hough transform.

The vascular structure of the retina can be affected when a person suffers some type of pathology, such as myopia, hypertension, diabetes, atherosclerosis and retinopathy of prematurity (ROP), with the blood vessels undergoing changes in their width, shape and tortuosity. Now, specifically for MTA, a decrease in the angle of insertion has been detected in at least two types of diseases: severe myopia and ROP [1]. Although there is no standard definition, the angle of insertion of the MTA is easy to understand as the angle formed by the upper and the lower parts of the MTA, known as the superior temporal arcade (STA) and inferior temporal arcade (ITA), both starting from optic nerve head and continuing its extension to the periphery of the retina. This angle is also known as the arcade angle, and it allows evaluating the structural integrity of the macular region [2, 3].

Fledelius and Goldschmidt [4] carried out a study related to the changes in geometry that the MTA presents specifically for the patients with myopia, finding that the decrease in the insertion angle was directly related to the severity of the patients' myopia. The measurements were made manually by ophthalmologists who located the MTA, and by placing the strategic points, lines were drawn, and in this way, the opening angle was measured.

However, it is important to be able to optimize the detection of the MTA. In the literature, the Hough transform is the most prominent method for the detection of parametric objects [5–7], this being a standard technique for shape recognition, being very useful in image analysis. *Oloumi et al.* [1] proposed to parametrize the MTA using a parabolic model. Using a Gabor filter bank, the vascular structure was detected to subsequently apply the Hough transform to adjust a parabola to the skeletonized section of the vascular structure.

DOI: 10.1201/9781003373117-6

But, the MTA is not exactly a parabola, even when the image is obtained from a healthy patient. This is why in a subsequent work by *Oloumi et al.* [3], the proposal is based on placing two parabolas that fit the MTA, with the intention of solving the non-symmetry problem. Although the Hough transform is the most used technique reported in the literature for detecting geometric objects, its great disadvantage is the computational time used, since it uses the exhaustive search strategy.

Due to this, new methods for modeling parabolas have recently appeared in the literature that significantly overcome the computational time used by the Hough transform. *Valdez et al.* [8], using a fast hybrid method that combined the Univariate Marginal Distribution Algorithm (UMDA) distribution estimation algorithm with the simulated annealing, were able to detect parabolas in order to locate the MTA in fundus images, besides other evolutionary algorithms such as those presented by *Guerrero-Turrubiates et al.* and *Jaime Giacinti et al.* [9, 10]. Since it is sought to obtain solutions in real time to be applied in clinical practice, these methods are supposed to be promising.

In this work, a method is proposed that performs the modeling of the MTA by means of a fifth-order polynomial for previously segmented fundus images from the DRIVE database [11]; the coefficients of the polynomials are determined by means of two evolutionary algorithms, besides the genetic and the quantum genetic algorithms. Before modeling, the image goes through pre-processing where the vein is skeletonized. Subsequently, the genetic algorithm and the quantum genetic algorithm are executed, each following the steps that will be described in the subsequent sections.

In order to evaluate the efficiency of the method, the computational time used is measured, as well as how good the fit is on the set of the skeletonized pixels. Besides, the experimental results are contrasted with several existing methods in the state of the art. As a consequence of the experimental results, the proposed method is capable of numerically modeling the MTA with good precision and low computational time, making the method as a good candidate for use in the medical practice of ophthalmology.

The chapter is organized as follows: in Section 6.1, a review of the state of the art has been provided; Section 6.2 gives a description of the classical polynomial fitting, and it also analyzes in detail how the genetic and quantum genetic algorithms work and the explanation of the proposed method and the types of the metrics used to verify the results; Section 6.3 shows the computational experiments using the database DRIVE and the discussion of the results; Section 6.4 presents the principal and relevant conclusions obtained from this chapter.

6.2 BACKGROUND

This section introduces the general fundamentals of polynomial fitting as well as the description of two evolutionary algorithms, the genetic algorithm and the quantum genetic algorithm, addressing their respective implementation methodologies.

6.2.1 Database of Major Temporal Arcade Images

The publicly available DRIVE database of 40 retinal fundus images has been used to perform the numerical modeling of the Major Temporal Arcade. The retinal images are of size 565×584 pixels and RGB format of 8 bits per color plane.

Since the DRIVE database is used for blood vessel segmentation, it was analyzed by an expert ophthalmologist at the High Specialty Medical Unit UMAE-T1 from the Mexican Social Security Institute, in order to specifically delineate the Major Temporal Arcade. The database was divided into the subsets of training and testing, with 50% and 50% of images, respectively.

6.2.2 Polynomial Fitting

Curve fitting is an optimization problem in which a curve is sought to approximate as closely as possible to a collection of data. This adjustment involves first defining the functional form of the mapping function and then searching the parameters (coefficients) of the function that result in the minimum error [12].

Let there be a set of data given as $\{(x_i, y_i)\}_{i=0}^{N}$ from which it is sought to find a functional approximation $f(x)$; it is possible to do by means of a polynomial function $p(x)$ that interpolates the data, which will happen if the relation $p(x) = y_i$, $i = 0, 1, 2, \ldots, N$ is satisfied. A system of $N+1$ equations will be created, which covers the interpolation conditions. There are several functional forms for $p(x)$; however, polynomials, as there are efficient methods to determine and evaluate them, become excellent candidates. The polynomial that interpolates the data set is known as the interpolating polynomial.

In general, there is an infinite number of polynomials that interpolate a data set, by simplicity, let be a power series polynomial:

$$p_M(x) = a_0 + a_1 x \mp a_2 x^2 + \ldots + a_M x^M \tag{6.1}$$

The last equation is of M degree; if that polynomial interpolates the set $\{(x_i, y_i)\}_{i=0}^{N}$, a linear system with the interpolation conditions is formed as

$$p_M(x_0) = a_0 + a_1 x_0 + a_2 x_0^2 + \ldots + a_M x_0^M = f_0$$
$$p_M(x_1) = a_0 + a_1 x_1 + a_2 x_1^2 + \ldots + a_M x_1^M = f_1$$
$$\vdots$$
$$p_M(x_N) = a_0 + a_1 x_N + a_2 x_N^2 + \ldots + a_M x_N^M = f_N$$

The previous linear system will have a unique solution if and only if the system is square ($M = N$). The matrix of coefficients of the linear system of interpolating conditions is known as the Vandermonde matrix:

$$
\begin{pmatrix}
1 & x_0 & x_0^2 & \cdots & x_0^M \\
1 & x_1 & x_1^2 & & x_1^M \\
& \vdots & & \ddots & \vdots \\
1 & x_N & x_N^2 & \cdots & x_N^M
\end{pmatrix}
\tag{6.2}
$$

Through some mathematical calculations, it can be shown that the determinant of the previous matrix is given as:

$$
\det V_N = \prod_{i>j}\left(x_i - x_j\right)
\tag{6.3}
$$

and since the determinant is non-zero, it can be concluded that the matrix V_N is non-singular, if and only if the nodes $\{x_i\}_{i=0}^{N}$ are different.

In order to determine the interpolating polynomial, it is necessary to find the coefficients of the power series that make it up, for this is necessary to solve the linear system formed by the pairs of points that make up the database $\{(x_i, y_i)\}_{i=0}^{N}$:

$$
V_N C = f_i
\tag{6.4}
$$

where C corresponds to the coefficient vector, V_N is the Vandermonde matrix and f_i is the y_i vector values from each ordered pair. However, the number of calculations to solve the linear system grows as N^3, N being the degree of the polynomial. The computational cost can be significantly high, but this could be reduced if properties of the Vandermonde matrix are exploited or if the representation of the interpolating polynomial is considered to be modified.

Choosing the way of modifying the polynomial representation, there are several well-established techniques; for example, the Newton's form, the Lagrange's form, Hermite interpolation, interpolation by splines, Chebyshev polynomials and so on.

Once the N-degree interpolating polynomial that fits the data set has been found, it is important to know how well this polynomial approximates the function $f(x)$ at any point x. Let the evaluation point x and the whole set points $\{x_i\}_{i=0}^{N}$ be quantities that lie in a closed interval $[a,b]$; it is possible to show by Rolle's Theorem that the error can be written in the form

$$
f(x) - p_N(x) = \frac{\omega_{N+1}(x)}{(N+1)!} f^{(N+1)}(\xi_x)
\tag{6.5}
$$

where $\omega_{N+1}(x) \equiv (x - x_0)(x - x_1) \cdots (x - x_N)$ and $\xi(x)$ is some point in the closed interval $[a,b]$. The specific location of the point depends largely on $f(x)$, x and $\{x_i\}_{i=0}^N$. The following are some of the properties of the interpolation error:

- The error will be zero when $x = x_i$; this is because in the fraction, the numerator will be $\omega_{N+1}(x) = 0$.
- The error is zero when f_i is a measurement of a polynomial $f(x)$ of degree N. This comes directly from the uniqueness theorem of polynomial interpolation.

Take the absolute value of each side of Equation 6.5, and further maximizing both sides, the polynomial interpolation error limit is obtained:

$$\max_{x \in [a,b]} |f(x) - p_N(x)| \leq \max_{x \in [a,b]} |\omega_{N+1}(x)| \cdot \frac{\max_{z \in [a,b]} |f^{(N+1)}|}{(N+1)!} \tag{6.6}$$

In order to get a small error, one must look for one of the terms $\dfrac{\max_{z \in [a,b]} |f^{(N+1)}|}{(N+1)!}$ or $\max_{x \in [a,b]} |\omega_{N+1}(x)|$ to be small. However, it is more common to have some information regarding $f(x)$ or its derivatives, so it is difficult to minimize the expression that contains the derivatives. Thus, attention should be directed to the nodes $\{x_i\}_{i=0}^N$ such that $|\omega_{N+1}(x)|$ should be small through $[a,b]$.

Although the method described here is widely used in the problems involving curve fitting, the method may face situations where it becomes very difficult or even impossible to determine the coefficients of the polynomial. Taking this into consideration, two evolutionary algorithms that can overcome this limit and solve the problem are described next.

6.2.3 Genetic Algorithms

Genetic Algorithms (GA) are a stochastic optimization strategy, which uses a number of potential solutions to address a complex problem that would be very difficult to solve using classical (deterministic) methodologies. The GAs are based on the principles proposed by Charles Darwin: the "survival of the fittest" [13, 14]. In 1975, Holland visualized the concept, which was able to culminate in a publication [15].

In general, a GA starts with a random population of individuals (solutions), which is usually encoded in a binary string. The population is evaluated using a fitness value, which is directly related to the function of the optimization problem. The initial population is evolved through the natural genetic operators, which are crossover, mutation and selection. This is repeated iteratively until a stopping criterion is reached.

As mentioned earlier, GAs work with encoded variables. This is a great advantage since it allows working with a discretized search space even when the function is continuous. Also, in contrast to traditional optimization methods, which work with one solution, GA uses a population of potential solutions at a time. Namely, GAs, being an iterative optimization technique, have a space of solutions, referred to as population in each generation.

The genetic operators are applied in each iteration, and the number of iterations in the algorithm indicates the number of generations that will be performed on the population. Next, the basic vocabulary used in the GAs is explained:

- Population: At the beginning of the algorithm, a search space is initialized randomly, where each element of the population is known as an individual (potential solutions). New populations will be generated from which the best solution will be extracted when the stopping criterion is reached.
- Chromosome: The elements that make up a population are called individuals or chromosomes.
- Gene: Each element that makes up a chromosome is known as a gene. In a binary coding string, genes can take values of 0 or 1.
- Parent: Within the entire population, in each iteration or generation, candidates will be chosen to generate the offspring for the next generation.
- Child: New candidates generated from the parents of the previous generation.

In order to visualize and better understand the elements listed here, Figure 6.1 shows a graphic representation of them.

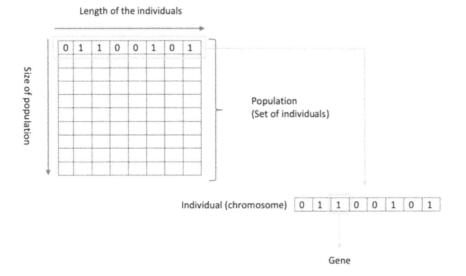

Figure 6.1 Schematization of the elements that make up a population of solutions.

On the contrary, the genetic operators are described next:

1. Selection operator: Depending on the objective of the problem, the most common strategy consists of ordering the individuals according to the fitness value, such that the most suitable are chosen to continue in the next generation.
2. Crossover operator: It consists of selecting a section of genes from two parents that will be exchanged, resulting in the creation of a new individual (offspring). Cross of one or several points can be performed. In general, it is established that a percentage of the population (60%) will be selected to be a part of the next generation. See Figure 6.2.
3. Mutation operator: It makes a change from 0 to 1 and vice versa in some individuals; a small mutation probability p_m must be considered. See Figure 6.3.

Next, the main idea behind the procedure performed by the GA is described next.

First, a population of n individuals, each one with a desired length, is initialized. This set of n individuals will form the initial population of possible solutions to the problem to be solved. Each individual is made up of a string of randomly chosen 0s and 1s, that is, a binary string. After the population is created, it is evaluated by a fitness function, in order to assign a fitness value for each individual. Now, the aim is to improve the population, and this is achieved by discarding the least fit individuals, for this is necessary to perform a ranking and make a selection of a part of the population. In general, between 60% and 70% of the individuals who will serve for the next generation are chosen.

Offspring must be created from the parents in the population; to do this, characteristics (genes) will be chosen that will be exchanged between them to form a new child individual, and this is achieved by applying the crossover operator. In order to maintain a diversity of the population, a quantity

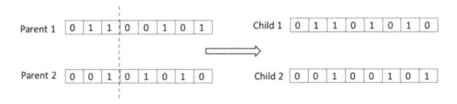

Figure 6.2 Crossover operator using the one-point strategy.

Figure 6.3 Mutation process of an individual.

of genes of certain individuals is changed from 0 to 1 or vice versa; this is called mutation and must be kept at a low percentage ($p_m = 5\%$). Mutation is useful for the local improvement of a solution.

After the previous steps are finished, a generation of the GA is completed. The process is repeated until a stopping condition is reached, generally when a number of generations (iterations) are finished. Figure 6.4 shows a block diagram of the process to follow the algorithm.

As previously mentioned, it is possible to perform curve fitting using an interpolating polynomial, whose coefficients are obtained by solving a system of linear equations whose matrix to invert is called the Vandermonde matrix. However, it was also mentioned that it may be the case that at the moment of building the matrix, a bad-conditioned situation appears, or the computation time required to make the adjustment is too long.

Thus, with the intention of avoiding these problems as well as showing the potential that evolutionary algorithms have in the optimization process, a GA was implemented to obtain the coefficients of an interpolating

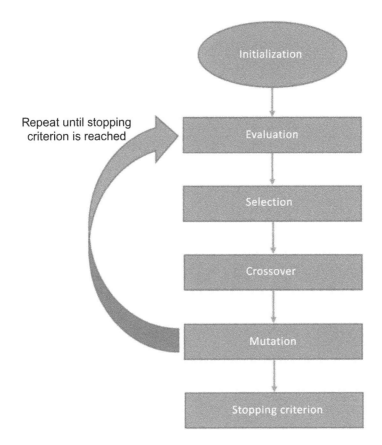

Figure 6.4 Block diagram of the GA.

polynomial that performs the best fit to a set of points that make up the Major Temporal Arcade (MTA) extracted for the fundus images.

In order to determine the interpolating polynomial, it is first necessary to pre-process the images. Next, the extraction of the coordinates of the pixels that make up the MTA is performed, and thus, the original data matrix is built.

It is time to start with the GA; a population of individuals is initialized, which will later be evolved through the crossover and mutation operators in such a way that when the fitness value is evaluated, it is directed toward the best optimal value that, according to the number of generations, goes forward, as is schematized in the block diagram for the GA. The Euclidean norm is used to calculate the fitness value of each chromosome of the population. Figure 6.5 shows the steps to follow all the process.

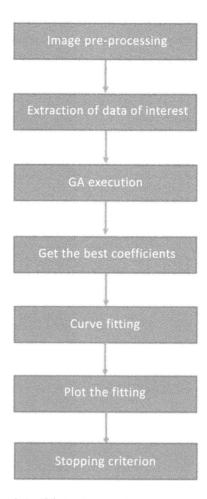

Figure 6.5 Flowchart of the GA implementation.

The GA was configured to initialize a population of 100 individuals, 200 generations to iterate and crossover and mutation percentages of 0.5 and 0.05, respectively.

6.2.4 Quantum Genetic Algorithm

With the intention to join quantum computing and the GAs, a new evolutionary algorithm came on. *Narayanan and Moore* proposed the quantum genetic algorithms (QGAs), with the TSP problem successfully solved [16]. The first difference between the GA and the QGA is related with a high efficiency, as well as its fast convergence, a very good global optimization and robustness [17–20]. Due to all of these, better results are expected through the QGA.

As the name suggests, the word "quantum" emphasizes the fact that some quantum mechanics concepts must be used. The principal, the quantum bits or simply "qubit", is defined as the smaller unit of information in a quantum evolutionary algorithm [21]. Just like in quantum mechanics, the quantum states will be the entities of work; separately, it is possible to start with the states "0" and "1", so the superposition of them will create the qubit, that is, $|\psi\rangle = \alpha |0\rangle + \beta |1\rangle$, the linear combination of the two states, being α and β the probability amplitudes from each state, which must satisfy the next condition $|\alpha|^2 + |\beta|^2 = 1$, with $(\alpha, \beta \in \mathbb{C})$. The qubit representation is

$$|\psi\rangle = \begin{pmatrix} \alpha \\ \beta \end{pmatrix} \tag{6.7}$$

so the "0" and "1" states' representation in a qubit form is

$$|0\rangle = \begin{pmatrix} 1 \\ 0 \end{pmatrix} \tag{6.8}$$

$$|1\rangle = \begin{pmatrix} 0 \\ 1 \end{pmatrix} \tag{6.9}$$

A better way to understand the concept of qubit is through a graphic representation, which allows assimilating and visualizing the operation of a qubit. The Bloch sphere is responsible as Figure 6.6 shows.

Now a population is required, and this can be constructed using the probability amplitudes of m qubits; such a quantity is named as the quantum population:

$$Q(t) = \begin{bmatrix} \alpha_1^t & \alpha_2^t & \cdots & \alpha_m^t \\ \beta_1^t & \beta_2^t & \cdots & \beta_m^t \end{bmatrix} \tag{6.10}$$

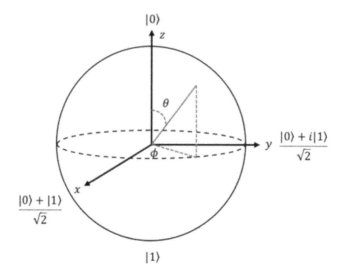

Figure 6.6 Bloch sphere for graphic representation of a qubit. Any point on the sphere will be a linear combination of the states $|0\rangle$ and $|1\rangle$. As examples, the points on the x and y axes are shown.

It is important to keep in mind that the population will evolve over time, and by convention, it is established that at $t = 0$, all probability amplitudes will set at $\dfrac{1}{\sqrt{2}}$, thus ensuring that all possible states start with the same probability.

In the previous paragraph, it was mentioned that the population must evolve over time, since all is about the quantum states, and the only way to achieve it is through the Schrödinger equation:

$$i\hbar \frac{\partial}{\partial t}\big|\psi\left(t\right)\big\rangle = H\left(t\right)\big|\psi\left(t\right)\big\rangle \tag{6.11}$$

where $i = \sqrt{-1}, \hbar = \dfrac{h}{2\pi}, H\left(t\right)$, the Hamiltonian of the system, and $|\psi\left(t\right)\rangle$ is a vector (wave function), which describes the state of quantum system in time. When an expression for the Hamiltonian is known, the equation can be solved; however, this operator will be understood as a sequence of operations performed on an initial state in a quantum computer.

Let's start by taking $U\left(t\right)$ as a solution to the Schrödinger equation, and then when it is applied to an initial state $|0\rangle$, its time evolution will be given as

$$\big|\psi\left(t\right)\big\rangle = U\left(t\right)\big|\psi\left(t\right)\big\rangle \tag{6.12}$$

Carrying out the computational analogy, let $|\psi(0)\rangle$ be the input of a computer, and therefore, $|\psi(t)\rangle$ would be the output as a consequence of the Schrödinger equation and a measurement made [22]; the latter will simply be the observation of the states of the qubits, and this process in physics is known as the collapse of the wave function. In more earthly words, simply when making the observation, the qubit will no longer be a linear superposition of both states ("0" and "1") but must take only one of them. On the contrary, the operator $U(t)$ is called in the QGAs as a quantum gate (Q-gate), and mathematically, it is simply a unit evolutionary operator.

In the summary form for the quantum genetic form, the following points are stated:

- Unlike a GA in this new algorithm (QGA), a new representation of the population of individuals is adopted, which provides a linear superposition of multiple states probabilistically.
- Having the population (possible solutions), through the Q-gate, it must be evolved in order to direct it toward more optimal individuals (better solutions). In this way, the individuals of the next generation will also be generated.

Several types of Q-gates can be found in the literature; one of the most important is explained next.

Rotation Gate: It is an operator that acts through unitary transformations and can be represented in a matrix form as

$$U(t) = \begin{pmatrix} \cos(\delta\theta_j) & -\sin(\delta_j) \\ \sin(\delta_j) & \cos(\delta_j) \end{pmatrix}. \qquad (6.13)$$

The unitary transformation will lead the state as

$$\begin{pmatrix} \alpha_j^{t+1} \\ \beta_j^{t+1} \end{pmatrix} = \begin{pmatrix} \cos(\delta\theta_j) & -\sin(\delta_j) \\ \sin(\delta_j) & \cos(\delta_j) \end{pmatrix} \begin{pmatrix} \alpha_j^t \\ \alpha_j^t \end{pmatrix}. \qquad (6.14)$$

It is observed that the matrix involves a rotation angle δ_j, which is already tabulated and is only searched for. This quantum genetic operator aims to modify (increase or decrease) the probability amplitudes of the qubits in such a way that the chromosomes of the population get closer to the chromosome with the maximum fitness value. A graphic representation of the process to a better understanding is shown in Figure 6.7.

Finally, a block diagram is presented in Figure 6.8 that describes the steps to be followed by a quantum genetic algorithm:

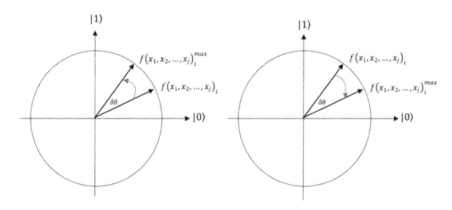

Figure 6.7 Graphic interpretation of the quantum genetic operator, Rotation Gate.

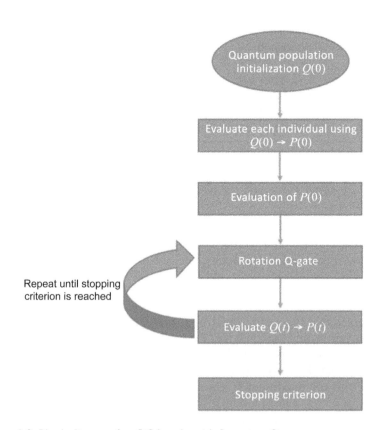

Figure 6.8 Block diagram for QGA only with Rotation Gate.

6.2.4.1 Implementation Details

Evolutionary algorithms of the QGA type continue to be implemented in the solution of optimization problems. In a recent work, *Hilali-Jaghdam et al.* [23] used the QGA for multilevel segmentation of medical images, where the algorithm is used to efficiently maximize Rèyni, Masi and Shannon entropies.

The quantum genetic algorithm is used to determine the coefficients of a polynomial of the form $P(x) = \sum_{i=0}^{M} C_i x^M$, whose objective is to adjust the pixels belonging to the MTA in fundus images.

As described in the last section, the algorithm presented begins by creating a population of individuals (possible solutions) represented by qubits, to measure and evolve later using the Q-gate evolutionary quantum genetic operator, which will direct the chromosomes to have better fitness values.

Figure 6.9 Flowchart of the QGA implementation.

This operator also guarantees to achieve diversity in possible solutions and mainly avoids stagnation in local minima within the search space.

The analyzed images have been previously outlined by an ophthalmological expert; each one is read and skeletonized in order to perform the extraction of the necessary pixels that allow the optimal adjustment to the MTA.

Once the pixel coordinate matrix is built, the QGA is executed to find the best coefficients that build up the equation of the polynomial that produces the best fit to the data. The Euclidean norm is used to calculate the fitness value of each chromosome of the population. Figure 6.9 shows the steps to follow.

Just like the GA, the QGA was configured to initialize a population of 100 individuals, 200 generations to iterate and a Q-gate as an evolutionary operator. Thus, it is possible to compare both algorithms.

6.2.5 Proposed Method

This section describes the proposed method to perform the modeling of the MTA in fundus images using a QGA evolutionary algorithm methodology. As previously mentioned, in the literature, several models are presented using parabolic approximations; however, as shown in Figure 6.10, the MTA is not a symmetrical parabola even when the image has been obtained from a healthy patient, and then in patients who present some ocular disease, the difference will be bigger.

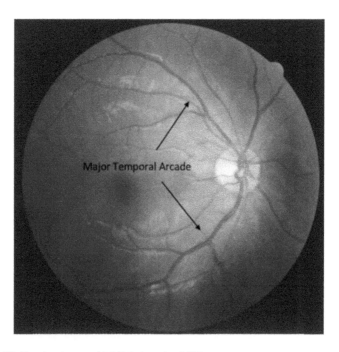

Figure 6.10 Fundus image, highlighting the MTA.

It is also known that the polynomials are considered good candidates to fit data sets [24–26]; taking this into consideration in this work, it is proposed to use a polynomial in the form of a power series:

$$P(x) = \sum_{i=0}^{n} c_i x^i = C_0 + C_1 x + C_2 x^2 + \cdots + C_n x^n, \tag{6.15}$$

where, applying the evolutionary algorithm, the best set of coefficients C_i $(i = 0, 1, 2, \ldots, n)$ that provide the polynomial with the best fit to the MTA pixels is found.

The evolutionary methodology used was the QGA, which uses a population of possible solutions made up of qubits, which, after an observation that transforms every individual to a binary chain, is evolved through a quantum genetic operator, referred to as the Q-gate. The individuals who make up the population are evaluated at each evolutionary step, and those with the best fitness value are selected to be used as a part of the new population that will be evolved and measured in the next generation.

After having carried out several experiments, it was determined that the most suitable polynomial expression to model the MTA is a fifth-order polynomial. That said, the algorithm aims to find the best six coefficients that determine it.

Once the stopping criterion is reached, which, for this case, consists of finishing the number of iterations (generations) setting in the initial configuration, the coefficients were evaluated in the polynomial function and were plotted together with the ground-truth. Metrics such as Mean Distance to the Closest Point (MDCP) and the Hausdorff distance were calculated in order to determine how good the polynomial fitting is obtained by the QGA.

The most representative steps of the proposed methodology for modeling the MTA are presented in Algorithm 1.

Algorithm 1 MTA modeling by Quantum Genetic Algorithm

Input: Population Size, Length Chromosomes, Generations
Output: P_{best}
1. *Initialize Quantum Population (Population Size, Length Chromosome)*
2. *Observe Population (Q-Population)*
3. *Evaluate Population (Population)*
4. *Get Best Solution*
5. **For** *gen 2 to Generations* **do**
6. *Observe Population (Q-Population)*
7. *Evaluate Population (Population)*
8. *Quantum Gate (Q-pop, Pop, fitness, BestChromosome)*
9. *Get New Best Fitness*
10. **if** *New Best Fitness is better than Best Fitness* **then**
11. *Upload Best Solution*
12. **end if**
13. **end for**
14. **return** P_{best}

Table 6.1 Example of an Individual; Six Coefficients Represented with Each One
10 Bits.

Individual		
C_0	C_1	...
0011011001	101011000	...

The length of each chromosome in the population corresponds to 60 bits, taking 10 bits per evaluated coefficient. Table 6.1 demonstrates the same.

6.2.6 Evaluation Measures

Once the polynomial that best fits the data set has been determined, it is necessary to measure its closeness with respect to these. There are several measures in the literature that allow these measurements to be made; however, in order to maintain the same ones that were used in the work of Oloumi *et al.* [3], MDCP and Hausdorff distance were used. A brief description of each of them is presented next. For MDCP, there are two sets of points A and B. The point of the approximation is in A, and the original data are in B; thus, MDCP is defined as:

$$MDCP(A,B) = \frac{1}{N}\sum_{i=1}^{N} DCP(a_i, B), \qquad (6.16)$$

where N is the cardinality of A, a_i is its ith element and DCP is the distance to the closest point, which can be calculated as follows:

$$DCP(a_i, B) = \min \|a_i - b_j\|, \qquad (6.17)$$

with $j = 1, 2, \ldots, M$, M being the cardinality of B, its jth element b_j and $\|\cdot\|$ a norm operator, which is generally chosen as the Euclidean norm. The MDCP measures an average of the distance for each point of one set to the closest point of the other set.

Hausdorff distance performs a similar calculation to the previous one, that is, the part of the definition of DCP is the same, and it is necessary to find the distance from each point of set A to the elements belonging to set B; however, instead of calculating an average, it takes the maximum value:

$$H(A,B) = \max DCP(a_i, B). \qquad (6.18)$$

Once both measures have been calculated, small values are indicative that the model is a good fit for the ground-truth.

6.3 COMPUTATIONAL EXPERIMENTS

All experiments in this study were carried out with Intel®Core™i5–3210M CPU @ 2.5–3.1 GHz and 8 GB RAM. The algorithms were coded and evaluated in MATLAB®R2021b running on MacOS Catalina. Using the set of fundus images from the DRIVE database, 30 executions of both GA and QGA were performed for each image, in order to have sufficient information to perform an adequate statistical analysis.

In Table 6.2, a comparison between the proposed method values and those reported in literature is shown about the MDCP and Hausdorff distances. For the MDCP, it is observed that the smallest distances were obtained with the GA and QGA algorithms, where even the QGA presents a difference of 13 pixels with respect to the GA, which would be the next method with the lowest value. The Hausdorff distance was the method that presented the largest measurement; however, the QGA was again positioned as the one with the smallest distance with a value very close to that obtained by the Medical Image Processing, Analysis and Visualization (MIPAV) method.

Table 6.3 shows the execution time that each method takes. The shortest time is obtained by the UMDA+SA method with 1.68 seconds; however, the GA and QGA algorithms also offer a short execution time, with QGA being 3.18 seconds faster than GA.

It is important to mention that the MIPAV tool is a free software, and it has been investigated that the internal algorithm for calculating the parabola is

Table 6.2 Different Methods' Comparison about the Closeness between Modeling Approximation and Ground-truth MTA Delineation. Average of 30 Runs of the MDCP and Hausdorff Distances.

Method	MDCP (px.) Mean±Std.	Hausdorff (px.) Mean±Std.
General Hough	31.25 ± 0.00	64.49 ± 0.00
MIPAV	25.69 ± 0.00	59.91 ± 0.00
UMDA+SA	30.45 ± 12.94	105.8 ± 27.54
GA	25.47 ± 7.89	129.82 ± 68.13
QGA	12.04 ± 3.94	59.18 ± 27.48

Table 6.3: Execution Time Comparison of MTA for Modeling Using the Data Set Test.

Method	Execution time (s)
General Hough	4.7641 (per pixel)
MIPAV	230
UMDA+SA	1.68
GA	7.82 ± 0.11
QGA	4.64 ± 0.33

based on an implementation of the Hough transform, restricting the degrees of freedom for its rotation. The performance of the Hough transform is often affected by the language of the programming and optimization of the developed code. That's why this is the high execution time for the MIPAV method.

Tables 6.4 and 6.5 show the maximum and minimum value, variance, median and mean for the 30 iterations of the best solution found with the GA and QGA, respectively.

Figures 6.11 and 6.12 show the curve fitting obtained by the GA and QGA, respectively; a strategic selection of 10 images was made. For each image, the MTA is shown in white, while the curve fitting is shown in red. When observing in detail, it is possible to note that the curve fitting obtained through the QGA presents a better behavior than that given by the GA. The curve generated by the polynomial obtained by the QGA is better adapted to the MTA. Considering this and also that it has a shorter execution time as well as a shortest distance in the MDCP and the Hausdorff distance, the QGA becomes an excellent method for its implementation in the numerical modeling of the MTA.

Regarding the general equations for each polynomial that models the MTA, Table 6.6 shows the functional form of each one. It is easy to notice that there are substantial differences in the coefficients that accompany each power of x of the polynomial, whether they are obtained by the GA method or the QGA method.

Table 6.4 Statistical Values for the Best Solution of the GA.

	MDCP (px.)	Hausdorff (px.)	Time (s)
Maximum	39.87	289.06	8.19
Minimum	12.18	29.74	7.62
Variance	62.32	4642.53	0.01
Median	25.09	106.51	7.81
Mean	25.47	129.83	7.83

Table 6.5 Statistical Values for the Best Solution of the QGA.

	MDCP (px.)	Hausdorff (px.)	Time (s)
Maximum	19.78	110.69	6.16
Minimum	5.26	20.68	4.41
Variance	15.54	755.66	0.11
Median	12.06	53.41	4.53
Mean	12.04	59.18	4.64

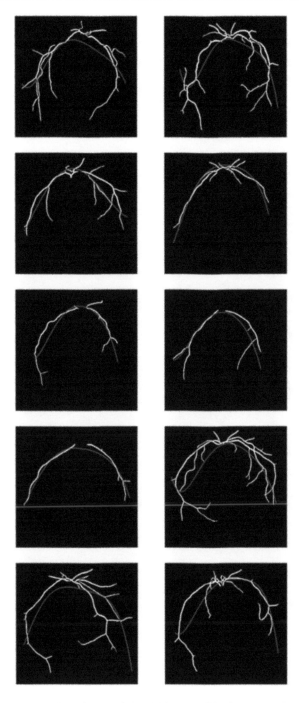

Figure 6.11 Polynomial modeling of the MTA using GAs for a set of testing images.

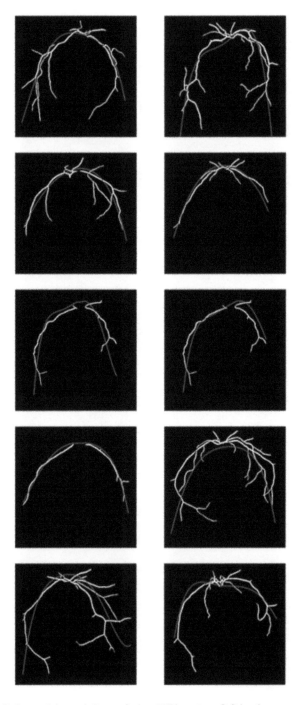

Figure 6.12 Polynomial modeling of the MTA using QGAs for a set of testing images.

Table 6.6 Fifth-order Polynomial Function from the Proposed Method.

Image	Type of algorithm	Function
01_test	GA	$20.08x^5 - 30.08x^4 - 47.96x^3 + 145.23x^2 - 20.57x + 112.36$
	QGA	$16.91x^5 + 9.71x^4 - 64.11x^3 + 11.69x^2 - 15.09x + 80.06$
03_test	GA	$41.70x^5 - 8.78x^4 - 99.32x^3 + 119.62x^2 + 18.62x + 108.07$
	QGA	$-0.85x^5 + 100.5x^4 + 97.25x^3 - 34.75x^2 - 113.75x + 109.88$
06_test	GA	$-0.72x^5 - 15.89x^4 - 24.02x^3 + 124.71x^2 + 67.93x + 82.95$
	QGA	$1.46x^5 + 11x^4 + 21.17x^3 + 87.64x^2 - 4.32x + 74.34$
07_test	GA	$4.76x^5 + 6.53x^4 - 22.44x^3 + 56.39x^2 - 14.45x + 58.34$
	QGA	$1.33x^5 + 3.74x^4 - 10.27x^3 + 60.48x^2 - 10.37x + 61.41$
09_test	GA	$20.09x^5 - 16.99x^4 - 16.99x^3 + 138.54x^2 - 70.53x + 84.7$
	QGA	$-83.4x^5 + 123.58x^4 + 135.58x^3 - 13.11x^2 - 78x + 75.13$
10_test	GA	$12.14x^5 - 4.66x^4 - 4.28x^3 + 125x^2 - 27.21x + 105.78$
	QGA	$-43.31x^5 + 98.46x^4 + 41.33x^3 - 33.44x^2 - 82.8x + 98.16$
12_test	GA	$15.88x^5 27.35x^4 - 28.25x^3 + 76.43x^2 + 19.86x + 100.96$
	QGA	$22.73x^5 + 44.72x^4 - 7.85x^3 + 79.77x^2 - 4.67x + 74.99$
14_test	GA	$25.81x^5 - 12.98x^4 - 46.81x^3 + 143.13x^2 - 2.34x + 80.92$
	QGA	$8.53x^5 + 38.81x^4 - 0.68x^3 + 55.13x^2 - 9.44x + 95.35$
16_test	GA	$8.72x^5 + 23.82x^4 - 23.59x^3 + 17.78x^2 + 30.47x + 119.27$
	QGA	$-17.55x^5 - 0.63x^4 + 37.11x^3 + 107.49x^2 + 52.21x + 58.56$
17_test	GA	$25.59x^5 - 3.82x^4 - 54.63x^3 + 90.67x^2 - 49.88x + 83.84$
	QGA	$-11.54x^5 + 36.29x^4 + 7.47x^3 - 25.51x^2 + 16.08x + 103.45$

6.4 CONCLUSION

In this work, a method has been proposed to model the MTA in fundus images using fifth-order polynomial equations. The method consists of determining through evolutionary algorithms the value of the coefficients for a fifth-order polynomial that best fits the coordinates of the pixels that make up the MTA. It begins by acquiring the information of the imaging that was previously segmented. A quick processing is applied to the image in order to extract the pixels of greatest interest; for this, the image is rotated 90° and skeletonized. Subsequently, the GA and QGA are executed with the intention of making a comparison regarding their performance. The similarity between the modeling obtained by the proposed method and the delineation of the vein made by an expert is then quantified by means of two different metrics, MDCP and Hausdorff distance, since both express measurements of closeness between the two sets of points.

According to the experimental results, it can be seen that the proposed method is better than the specialized methods found in the literature against which it is compared. It is observed that specifically the QGA is better than the GA, since both the MDCP and Hausdorff distance are less than half for the first compared with the second. Regarding the Hough transform, the QGA presents a difference of 19 pixels in the MDCP and 5 pixels in the Hausdorff distance. A similar difference occurs for the MIPAV and UMDA methods. As for the execution time, the QGA is 3.18 seconds faster than the GA; however, with respect to the other methods, the UMDA is still the fastest.

According to the results obtained and the comparisons made with the other methods discussed earlier, the proposed method can be considered as a good candidate to be used in systems that perform computer-assisted ophthalmological diagnosis.

Appendix

Matlab Code

In this Appendix, the source code in Matlab environment of the QGA for data fitting is provided.

```
1
2 % QGA CODE FOR DATA FIT USING POLYNOMIAL
3 %The algor i thm l o o k s f o r the be s t va lue s f o
  r the c o e f f i c i e n t s
  that wi l l use
4 %a polynomial func t i on o f the form P(x)=C0+C1*x+C2*x
  ^2+ . . . +C^N
5 %-----------------------------------------------------------
6 c l e a r a l l;
7 c l o s e a l l;
8 c l c;
9 t i n i c i o=t i c;
10 % Se t t ing the i n i t i a l parameter
11 n=50; %Number o f i n d i v i d u a l s
12 L=60; %Length o f the chromosomes
13 itmax=200; %Maximum number o f i t e r a t i o n s (g ene
   r a t i ons)
14 %-----------------------------------------------------------
15 %The data to be analyzed i s loaded (Image data)
16 I1=imrotat e (imc l e a rbo rde r (imread (' 01 Input/S i
   n g l e /MTA 12. png '))
   ,90);
17 I2=bwskel (I1);
18 [y, x]=f i n d (I2);
19 XY=[x y];
20 %-----------------------------------------------------------
21 [p, ~,mu]= p o l y f i t (x, y, 5);
22 minimo=min (p);
23 maximo=max(p);
24 rmax=redondeo (maximo);
25 rmin=redondeo (minimo);
26 %-----------------------------------------------------------
```

```
27 be s t=s t r u c t (' f t n s ',0, 'C', [], 'Qm', [],
   'Q', []);
28 di sp ('QUANTUM GENETIC ALGORITHM TO DATA FITTING');
29 f p r i n t f (' \n ')
30 Q=i n i c i a rPo b (2*n,L); %Tehe i n i t i a l popul a
   t i ons begins
31 Qm=c o l aps a r (Q); %The populat ion i s observed
32 [f tns, c o e f]=evaluar3 (Qm, x, y, rmax, rmin,mu); %The
   f i t n e s s value i s
   obtained
33 %[f tns, c o e f]=evaluar3 (Qm, x, y);
34 %Looking f o r the sma l l e s t f i t n e s s value
35 minimo=min (f t n s);
36 pos=f i n d (f t n s==minimo); %f i n d the p o s i t i o
   n o f the e be s t f i t n e s s
   value
37 %Now the c o e f f i c i e n t s that g ive t h i s f i t
   n e s s value ar e ext r a c t ed
38 bestCrom=Qm(pos,:);
39 be s tCoe f=c o e f (pos,:);
40 bestQ=Q(pos,:);
41 be s t. f t n s=minimo;
42 be s t. C=be s tCoe f;
43 be s t. Qm=bestCrom;
44 be s t. Q=bestQ;
45 %Q=Qgate (Q, f tns, bes t,Qm);
46 t r a c e (1)=be s t. f t n s;
47 f p r i n t f (' Generat ion %d\n ', 1)
48 %%%%THE EVOLUTION OF TE POPULATION BEGINS (EVOLUTION OF THE
   ALGORITHM)%%%%
49 f o r gen=2: itmax
50 f p r i n t f (' Generat ion %d\n ', gen)
51 Qm=c o l aps a r (Q); %the populat ion i s observed
52 [f tns, c o e f]=evaluar3 (Qm, x, y, rmax, rmin,mu); %Fi
   tne s s value i s
   c a l c u l a t e d
53 % The Quantum gat e i s appl i ed (quantum ope r a t o r)
54 Q=Qgate (Q, f tns, bes t,Qm);
55 newBestFtns=min(f t n s);
56 pos=f i n d (f t n==newBestFtns);
57 i f newBestFtns<be s t. f t n s
58 be s t. f t n s=newBestFtns;
59 be s t. C=c o e f (pos,:);
60 be s t. Qm=Qm(pos,:);
61 be s t. Q=Q(pos,:);
62 end
63 t r a c e (gen)=be s t. f t n s;
64 end
65 %The v o lut i ona r y curve i s drawn
66 f i g u r e (); pl o t (1: itmax, t r a c e); g r id;
```

```
67 t i t l e (' Evolut ion ');
68 x l a b e l (' Generat ion ');
69 y l a b e l (' Best Fi tne s s Value ');
70 %The r e s u l t i s pl o t t ed a f t e r the e v o lut
   i on o f the t o t a l number
   o f g ene r a t i ons
71 %The o r i g i n a l data need to be c ent e r ed and s
   c a l e d
72 xn=z e r o s (l eng th (x), 1);
73 f o r i =1: l eng th (x)
74 xn (i)=(x (i)=mu(1)). /mu(2);
75 end
76 %-----------------------------------------------------------
   ---------------------------------
77 % Polynomial func t i on
78 P=@(C) C(1) *xn.^5+C(2) *xn.^4+C(3) *xn.^3+C(4) *xn.
   ^2+C(5) *xn+C(6);
79 f i g u r e (); pl o t (x, y, ' k. ',x, po lyva l (be s
   t. C, x, [],mu), ' r='); g r id
80 %-----------------------------------------------------------
   ---------------------------------
81 l egend ({'Datos ', ' Ajus te '})
82 %The d i s t a n c e s MDCP and Hausdor f f ar e c a l c
   u l a t e d
83 newXY=[x P(be s t. C)];
84 %newXY=[xn po lyva l (be s t. C, x, [],mu)];
85 %newXY=[x P(be s t. C)];
86 %newXY=[x po lyva l (be s t. C, x, [],mu)];
87 mdcp=MDCP(XY,newXY);
88 H=hausdo r f f (XY,newXY);
89 f p r i n t f (' \nThe MDCP di s t anc e i s: %f \n
   ',mdcp);
90 f p r i n t f ('The Hausdor f f di s t anc e i s: %f \n
   ',H);
91 t f i n=toc (t i n i c i o);
92
93 %-----------------------------------------------------------
   ---------------------------------
94 % GA FOR DATA FITTING USING POLYNOMIAL
95 %One point c r o s s o v e r and mutation
96
97 c l e a r a l l
98 c l o s e a l l
99 c l c
100 t i n i c i o=t i c;
101 %-----------------------------------------------------------
    ---------------------------------
102 % Data s e t to f i t (Image p i x e l s)
103 I1=imrotat e (imc l e a rbo rde r (imread (' 01 Input/S
    i n g l e /MTA 12. png '))
    ,90);
104 I2=bwskel (I1);
```

```
105 [y, x]=f i n d (I2);
106 XY=[x y];
107 %--------------------------------------------------------
    ----------------------------------
108 [p, ~,mu]= p o l y f i t (x, y, 5);
109 minimo=min (p);
110 maximo=max(p);
111 rmax=redondeo (maximo);
112 rmin=redondeo (minimo);
113 %--------------------------------------------------------
    ----------------------------------
114 di sp ('GENETIC ALGORITHM ONE CROSSOVER POINT');
115 % I n i t i a l parameter s
116 n=100; % I n d i v i d u a l s
117 l =60; %l enght o f chromosome
118 itmax=200; % Number o f g ene r a t i ons
119 s =0.6; % S e l e c t i o n pe r c entage
120 cm=n / 2; %Pai r to chromosomes to c r o s s
121 pm=0.05; %Mutation pe r c entage
122 %--------------------------------------------------------
    ----------------------------------
123 be s t=s t r u c t (' f t n s ',0, ' c o e f ', [], '
    ind ', []);
124 % Cr eat ing i n i t i a l populat ion
125 P=inipob (n, l);
126 [f tns, c o e f]=eva luar (P, x, y, rmax, rmin,mu);
127 minimo=min (f t n s);
128 pos=f i n d (f t n s==minimo);
129 be s t. f t n s=minimo;
130 be s t. c o e f=c o e f (pos,:);
131 be s t. ind=P(pos,:);
132 t r a c e (1)=be s t. f t n s;
133 %Evolve the populat ion
134 f o r i =1: itmax
135 f p r i n t f (' Generat ion %d\n ', i);
136 C=cruza (P, cm, l); %Perform the c r o s s o v e r
137 M=mutacion (C, round (pm*n)); %Perform the mutation
138 [newftns, c o e f]=evalua r (M, x, y, rmax, rmin,mu);
139 newbe s t f tns=min (newf tns);
140 pos=f i n d (newf tns==newbe s t f tns);
141 Pnew=s e l e c c i o n (M, f tns, s); %Best i n d i v i
    d u a l s e l e c t i o n
142 i f newbe s t f tns<be s t. f t n s
143 be s t. f t n s=newbe s t f tns;
144 be s t. c o e f=c o e f (pos,:);
145 be s t. ind=P(pos,:);
146 end
147 P=Pnew; %The populat ion i s updated
148 t r a c e (i)=be s t. f t n s;
149 end
150 % The e v o lut i ona r y curve i s drawn
151 f i g u r e (); pl o t (1: itmax, t r a c e); g r id;
```

```
152 t i t l e (' Evolut ion ');
153 x l a b e l (' Generat ion ');
154 y l a b e l (' Best Fi tne s s Value ');
155 % The r e s u l t i s pl o t t ed a f t e r the e v o
    lut i on o f the t o t a l number
    o f g ene r a t i ons
156 f i g u r e (); pl o t (x, y, ' k. ',x, po lyva l (be s
    t. coe f, x, [],mu), ' r='); g r id
157 l egend ({'Data ', ' Fi t t i n g '})
158 %newXY=[x P(be s t. c o e f)];
159 newXY=[x po lyva l (be s t. coe f, x, [],mu)];
160 mdcp=MDCP(XY,newXY);
161 H=hausdo r f f (XY,newXY);
162 f p r i n t f (' \nThe MDCP di s t anc e i s: %f \n
    ',mdcp);
163 f p r i n t f ('The Hausdor f f di s t anc e i s: %f \n
    ',H);
164 t f i n=toc (t i n i c i o);
165 f p r i n t f ('The exe cut i on time i s: %f s e c \n
    ', t f i n);
166 t f i n=toc (t i n i c i o);
```

REFERENCES

[1] Oloumi F., Rangayyan R. M. and Ells A. L. Parabolic modeling of the major temporal arcade in retinal fundus images. IEEE Transactions on Instrumentation and Measurement. 61(7), 1825–1838, July 2021, doi:10.1109/TIM.2012.2192339.

[2] Wong K., Ng J., Ells A. L., Fielder A. R. and Wilson C. M. The temporal and nasal retinal arteriolar and venular angles in perform infants. British Journal of Ophthalmology. 95(12), 1723–1727, December 2011.

[3] Oloumi F., Rangayyan R. M. and Ells A. L. Dual parabolic modelling of the superior and the inferior temporal arcades in fundus images of the retina. 2011 IEEE International Symposium on Medical Measurements and Applications. 1–6, 2011, doi:10.1109/MeMeA.2011.5966784.

[4] Fledelius H. C. and Goldschmidt E. Optic disc appearance and retinal temporal vessel arcade geometry in high myopia as based on follow-up data over 38 years. Acta Ophthalmologica. 88(5), 514–520, August 2010.

[5] Ballard S. Generalizing the Hough transform to detect arbitrary shapes. Pattern Recognition. 13(2), 111–122, 1981.

[6] Illinworth J. and Kittler J. A survey of the Hough transform. Computer Vision, Graphics and Image Processing. 44(1), 87–116, 1993.

[7] Leavers V. Survey: Which Hough transform? Computer Vision Graphics and Image Processing: Image Understanding. 58, 250–264, 1993.

[8] Valdez S. I., Espinoza-Perez S., Cervantes-Sanchez F. and Cruz-Aceves I. Hybridization of the univariate marginal distribution algorithm with simulated annealing for parametric parabola detection. In: Bhattacharyya S. (ed) *Hybrid metaheuristics for image analysis*. Springer, Cham, 2018, doi:10.1007/978-3-319-77625-57.

[9] Guerrero-Turrubiates J. J., Cruz-Aceves I., Ledesma S., Sierra Hernandez J. M., Velasco J., Avina-Cervantes J. G., Avila-Garcia M. S., Rostro Gonzalez H. and Rojas-Laguna R. Fast parabola detection using estimation of distribution

algorithms. Computational and Mathematical Methods in Medicine. 2017, Article ID 6494390, 13 pages, 2017, doi:10.1155/2017/6494390.

[10] Jaime Giacinti D., Cervantes-Sanchez F., Cruz-Aceves I., Hernandez-Gonzalez M. A. and Lopez-Montero L. M. Determination of the parabola of the retinal vasculature using a segmentation computational algorithm. Nova Scientia. 11(23), 87–107, 2019, doi:10.21640/ns.v11i23.1902.

[11] Staal J., Abramoff M., Niemeijer M., Viergever M. and van Ginneken B. Ridge based vessel segmentation in color images of the retina. IEEE Transactions on Medical Imaging. 23, 501–509, 2004.

[12] Lam N. S. N. Spatial interpolation methods: A review. The American Cartographer. 10(2), 129–150, 1983.

[13] Gupta S. K. An overview of genetic algorithms: A structural analysis. International Journal of Innovative Science and Research Technology. 6(5), 1305–1309, 2021.

[14] Iquebal M. A. *Genetic algorithms and their applications: An overview*. White Paper, 2009, https://www.academia.edu/7140214/Genetic_Algorithms_Concepts_and_Applications.

[15] Holland J. H. *Adaptation in natural and artificial systems*. University of Michigan Press, Ann Arbor, 1975.

[16] Narayanan A. and Moore M. *Quantum-inspired genetic algorithms*. Proceedings of the IEEE International Conference on Evolutionary Computation (ICEC), Nagoya, Japan, pp. 61–66, May 1996.

[17] Wang H., Liu J., Zhi J. and Fu C. The improvement of Quantum genetic algorithm and its application on function optimization. Mathematical Problems in Engineering. 2013, Article ID 730749, 10 pages, 2013.

[18] Zhang G. Quantum-inspired evolutionary algorithms: A survey and empirical study. Journal of Heuristics. 17, 303–351, 2011.

[19] Sun Y. and Xiong H. Function optimization based on quantum genetic algorithm. Research Journal of Applied Science, Engineering and Technology. 7(1), 144–149, 2014.

[20] Roy U., Roy S. and Nayek S. Optimization with quantum genetic algorithm. International Journal of Computer Applications. 102(16), 1–7, 2014.

[21] Hey T. Quantum computing: An introduction. Computing and Control Engineering Journal. 10(3), 105–112, 1999.

[22] Lahoz-Beltra R. Quantum genetic algorithms for computer science. Computers. 5(4), 24, 2016, doi:10.3390/computers5040024.

[23] Hilali-Jaghdam I, Ishak A. B., Abdel-Khalek S. and Jamal A. Quantum and classical genetic algorithms for multilevel segmentation of medical images: A comparative study. Computer Communications. 162, 83–93, ISSN 0140-3664, 2020, doi:10.1016/j.comcom.2020.08.010.

[24] Choksi B., Venkitaraman A. and Mali S. Finding best fit for hand-drawn curves using polynomial regression. International Journal of Computer Applications. 174(5), 20–23, 2017.

[25] Tong Y., Yu L., Li S., Liu J., Quin H. and Li W. Polynomial fitting algorithm based on neural network. ASP Transactions on Pattern Recognition and Intelligent Systems. 1(1), 32–39, 2021.

[26] Ameer S. *Investigating polynomial fitting schemes for image compression*. University of Waterloo, Waterloo, ON, 2009.

Chapter 7

Quantum Logic Gate–Based Circuit Design for Computing Applications

Joy Bhattacharjee and Arpan Deyasi

7.1 INTRODUCTION

Quantum computing, which is the most famed topic in this era, a combination of quantum mechanics from the 1900s and the computing point of view, takes a huge place in the quantum realm. Starting from the foundation of abacus since 2700–2300 BC, we humans have always been trying to compute in different ways; the most modern way of doing it comes with the binary system. Bit is the smallest unit of a binary system and can be in two states, either 0 or 1. On the contrary, quantum computing deals with the "quantum bit" or "qubit"; possible outcomes from the qubits are neither 0 nor 1. It will always be a combination of 0 and 1, which is based on the superposition principle from quantum physics.

In this chapter, we will try to discuss the architecture of a quantum computer, the structure and operation principle of it, methods of quantum computing, and how it differs from the basic computing algorithm we used earlier. This chapter will also try to lead in the near future when quantum computing can be used for mutation of this world.

7.2 QUANTUM COMPUTING

Before understanding quantum computing, let's take an example. This example gives a basic idea about how quantum computing differs from the binary computation system. As the first example, we can take the flip the coin example. An anonymous user flips a coin. The outcome is usually either heads or tail. When the coin is in the air, it has also one state present at a instance of a time. This simple theory describes the binary computation system, where the outcome at any instance is always one [1]. Now we flip the coin again in the air while considering the quantum computing system. As per quantum computing, when the coin is in the air, it consists of a combination of heads and tail. In simple terms, the outcome from any instance in quantum computing theory is always more than one; it is a combination of two outcomes in a specific way [2].

A major difference with classical binary theory with quantum computing theory is that the alter uses qubit; qubits can exist with multiple outcomes at a single time using a theory called superposition. In the previous coin flipping example, we can say that the coin is flipping continuously, where the probabilities of both the existing head and tail are equal (50%). Due to this superposition, a qubit finds a solution to a problem with higher accuracy exponential times compared to the classical bit.

Using classical or Newtonian mechanics, every normal-sized object's position can be predicted accurately, but the problem comes around when we try to deal with subatomic particles moving at the speed of light, which gave rise to quantum mechanics and which deals with very small subatomic particles moving at a very high speed.

7.2.1 Superposition

In quantum systems, superposition is defined as two subatomic particles that can exist in two states at a single time until we observe it. This superposition is explained by an Irish Austrian physicist Erwin Schrödinger using a theoretical experiment referred to as the "Schrodinger Cat". In this experiment, Schrödinger stated that, if we place a cat and something that will kill the cat in a box and seal it, we cannot know if the cat is dead or alive until we open the box so that until the box is opened, the cat is (in a sense) both 'dead and alive' at the same time.

7.2.2 Quantum Entanglement

Quantum entanglement is when two particles are closely linked, and whatever happens to one particle immediately affects the other regardless how far apart they are. Albert Einstein named this behavior of quantum particles 'Spooky Behavior at a distance. This whole quantum entanglement theory is summarized as if two entangled particles are placed in very distant places like in two distant galaxies. In some mysterious way, one of the particle's states affects the state of the other particle and vice versa [3, 4].

7.2.3 Quantum Tunneling

From the basic physics, we all know that transistors combining emitter, base and collector work as the switch based on the current or voltage is applied to it. This switching behavior is the basic mechanism of modern computers using various logic gates. When this behavior comes to a quantum level, these rules of the basic semiconductor do not apply there. This is where quantum tunneling comes into picture. Quantum tunneling is the quantum mechanical phenomenon where subatomic particles travel from one side of the potential barrier and appear on the other side without any external energy applied inside the potential well. Due to this quantum tunneling, making switches using quantum scale particles becomes difficult.

7.3 QUANTUM BIT (QUBIT)

Classical bit is the very basic unit of classical computation. Classical bit consists of two values, 0 and 1. There are 2^n possible states, one at a time, where n is the number of bits. Classical bits can be created using transistors. Transistors are the units in an electrical circuit that act like a switch. Classical logic gates are made from the combination of transistors; these logic gates are grouped into modules that can perform bit-wise operation. Integrated circuits are made using embedding these gates [5]. Every character in the classical computer is converted to binary bits and is represented as it is.

Representation of bits in the quantum world is known as Qubit or quantum bit. Qubit is the basic unit of quantum information, similar to classical bit. Qubit also has states 0 and 1, but superposition, which we have already discussed in Section 1.3, allows the qubit to stay in both states 0 and 1 at the same instance of time [6]. There are 2^n possible states in classical bits with respect to qubit, where n is the number of qubits. We compare the classical bit with qubit in Table 7.1.

In Table 7.1, we compare classical and quantum bits with respect to their usage size. If the number of qubits is 3, then we can use 8 possible states at the same time; due to superposition, both 0 and 1 are available at the same instance. Similarly, for 10 qubits, 1024 bits and for 20 qubits, 1048576 bits can be accessed at the same time. This comparison clearly indicates how the qubit is faster and time efficient with respect to the classical bits. This leads to the minimum time requirement for the computation in a quantum computing system.

Classical bit can do a single computation at a given period of time, whereas due to the simultaneous existence of qubit in multiple states, qubits can do multiple calculations at the same time. The result of this multiple existence leads us to the parallel database search, or adaptive learning models used in artificial intelligence make it easier and faster compared with using classical bits.

Table 7.1 A Comparative Study between Classical and Quantum Bits

Quantum bit (Qubit)	Classical bit
3	8
10	1024
20	1048576

7.3.1 What Is Qubit?

Classical bits operated on the basis on the electron flow in between the transistors, whereas quantum bits or qubits operated on the basis on the spin of the electron or the orientation of a photon. In quantum computing, qubit's states are determined by the polarization or the orientation of a photon or an electron. The states of the qubits are determined by an electron's spin orientation, either up or down. Up-spin position is usually considered 1, and down-spin position is considered 0 state.

From the Pauli exclusion principle, we know the capacity or the highest number of electrons that can be accumulated in an orbit. Electrons rotate around the nucleus with a certain speed; the electrons also rotate upside down and vice versa during their rotation to the nucleus, which creates the spin up and spin down theory. Around the 1900s, many experiments were done to identify the spinning properties of electrons, with the most renowned experiment named the Stern-Gerlach Experiment. This experiment clearly indicates that electrons have spin movement with a certain speed [7]. In room temperature, we can't control the spin of the electrons due to the peripheral heat energy; as a result, in room temperature, electrons vibrate randomly without any controls. As a solution, these electrons cool down to a very low temperature, nearly absolute zero, to control the vibrations and the spin of those electrons. But direct transition from room temperature to the absolute zero temperature or 0K is not possible, so the temperature degradation follows a way like starting from 50K to 4K, 1K, 100mK, 50mK and 15mK. Reaching the absolute temperature is not possible till now in physical systems. The quantum processor is placed over the least temperature we can achieve. In this low temperature, electrons have a very low energy and become calm in a particular state. Then we can use the precision microwave to spin the electrons [8], but the electron's resonant frequency must be matched with the microwave frequency to make it vibrate and flip.

7.3.2 Formulation of a Qubit

In the previous section, we have defined the process how a qubit can be created. Now, we will discuss the qubit's state and its mathematical representation in detail.

When a qubit changing state, that is, spinning from 0 state to 1 state, did not happen in an instance, at a selected time period, the state of the qubit is undetermined due to the superposition; it will be a combination of 0 and 1 states. We are using a Dirac Bra Ket notation to determine the value of a qubit as a vector in a sphere:

$$< \Psi \mid = \alpha \mid 0 > + \beta \mid 1 >$$

where $\langle \Psi \mid$ represents the state of the qubit, $\mid 1 \rangle$ as up spin position and $\mid 0 \rangle$ as down spin position. $\pm$ and β are two arbitrary constants that determine the percentage or ratio of the qubit in each state.

Now, to determine the state of the qubit, for a time instance, we can assume the value for α_0 as 0.6 and α_1 as 0.4, where we take 1 or 100% as the margin. From these values of α_0, we can predict the probability of qubit in 0 state or 1 state; like in this case, the qubit is 60% probable to remain in the 1 state and 40% in the 0 state. This leads to the result that the qubit has a greater probability to be found out in 1 state when it is observed.

7.3.2.1 Creation and Retention of a QUBIT

As we know, in classical computers, the bits are stored in the memory register and can be retrieved without complexity. But in the case of qubit, these processes are not too easy like classical systems.

There are some natural qubits available, like C^{13} and Si^{29}, which possess the spin behavior in the presence of some external energy in the room temperature.

7.3.2.2 Phase in QUBIT

In general, the quantum state of a single qubit can be represented by the two-dimensional vector space over the complex number C^2. It means a qubit takes two complex numbers to completely describe it. $|\psi\rangle$, which is an arbitrary quantum state, can be represented as $|\psi\rangle = \alpha|0\rangle + \beta|1\rangle$, where α and β are two quantum amplitudes to determine the amount of superposition. By Born Rule [9], the result becomes 0 if this is obtained with probability $\backslash\alpha\backslash^2$ and the result becomes 1 with complement probability $\backslash\beta\backslash^2$. From the conservation of probability, we know that $\backslash\alpha\backslash^2 + \backslash\beta\backslash^2 = 1$, and since the global phase of a quantum state is not detectable, a single qubit state can be written as:

$$|\psi\rangle = \sqrt{(1-p)}\,|0\rangle + e^{i\phi}\sqrt{p}\,|1\rangle \tag{7.1}$$

where p is the probability of the qubit in high state or 1 state, $0 \le p \le 1$, that is, the value of p lies in between 0 and 1 and the value of ϕ lies in between 0 and 2π.

7.4 LOGIC GATES

Logic gates are the basic element for a big picture in real-life devices. Classical logic gates are based on transistors and are switching on the basis of current, as we discussed earlier. Quantum gates worked in a different way, unlike transistors in which qubits are the elements. In this section, we discuss different quantum gates such as Pauli X, Y and Z gates, Hadamard gate, R phi gate, S and T gates, U and I gates and so on [10].

Many popular organizations like Google, IBM, Intel and Microsoft made a number of frameworks to make quantum computing publicly available to use. All the gates mentioned here are single-qubit quantum gates that we discuss over next. As we know, we use voltage as the input variable in the classical gates; unlike that, we will be using electron spin as the input variable in quantum gates.

7.4.1 Pauli X, Y and Z Gates

Pauli X gate operation is similar to the NOT gate from the classical theory. The block diagram and the truth table of classical NOT gate is depicted in Figure 7.1 and Table 7.2, respectively.

The basic mechanism used in the NOT gate is converting the present state to its opposite state. A similar operation can be done using Pauli X gate through qubits [11]. As a result, if an electron has spin up as the input in the Pauli X gate, the output is that electron with spin down and vice versa. The matrix representation of the Pauli X gate is represented by the matrix:

$$\begin{bmatrix} 0 & 1 \\ 1 & 0 \end{bmatrix}$$

This matrix is called the Pauli X matrix or, generally, the transition matrix.

Pauli X matrix can be transformed into a state vector formation like:

$$|0><1|+|1><0|$$

The state vector of a qubit can be represented by two orthogonal vectors:

$$|0> = \begin{bmatrix} 1 \\ 0 \end{bmatrix} \tag{7.2}$$

$$|1> = \begin{bmatrix} 0 \\ 1 \end{bmatrix} \tag{7.3}$$

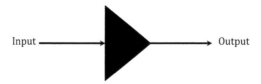

Input ⟶ Output

Figure 7.1 Block diagram of NOT gate.

Table 7.2 Truth table of NOT gate

Input	Output
High state or 1 state	Low state or 0 state
Low state or 0 state	Low state or 0 state

Now, we are taking an electron with low state, that is, |0>, or the spin up as the input; the output is obtained from the dot product of the input and transition matrix. The dot product is shown next:

$$\begin{bmatrix} 0 & 1 \\ 1 & 0 \end{bmatrix}\begin{bmatrix} 1 \\ 0 \end{bmatrix} = \begin{bmatrix} 0 \\ 1 \end{bmatrix} = |1> \tag{7.4}$$

In this equation, the dot product of the Pauli X matrix with the spin-up electron results in a spin-down electron. This perfectly matches with the algorithm of the classical NOT gate and the Pauli X gate in the quantum theory [12]. Similarly, with the spin-down electron as the input, the output we get from the Pauli X gate is the electron in the spin-up state. The equation is shown next:

$$\begin{bmatrix} 0 & 1 \\ 1 & 0 \end{bmatrix}\begin{bmatrix} 0 \\ 1 \end{bmatrix} = \begin{bmatrix} 1 \\ 0 \end{bmatrix} = |0> \tag{7.5}$$

So what we get from the Pauli X gate is the output that will be spin down if the input is spin up and the output that will be spin up if the input is the spin-down electron.

We have done the simulation of the Pauli X gate using IBM virtual quantum computer, and the circuit is shown in Figure 7.2. In this circuit simulation, we use a single qubit in a low state or an electron in the spin-up position [13].

The simulation results from Figure 7.2 with the probability distribution in Figure 7.3 where the maximum output probability is |1> state; we use 500 seeds to run the simulation.

Figure 7.2 Quantum circuit ssimulation of the Pauli X gate with a single qubit.

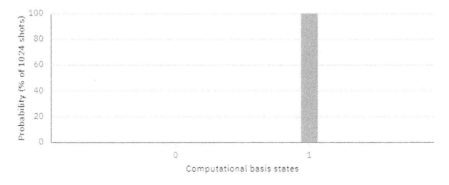

Figure 7.3 Probability distribution of the Pauli X gate with |0> state as the input.

In this context, we discuss about the Pauli Y gate, and the transition matrix in the Pauli Y gate is:

$$\begin{bmatrix} 0 & -i \\ i & 0 \end{bmatrix} \approx -i\,|\,0><1\,| + i\,|\,1><0\,|$$

where i is the imaginary number with the value of $\sqrt{-1}$. The translation matrix for Pauli Y gate has no difference from the Pauli X matrix other than this i parameter. In Pauli Y gate, electrons also rotate their spin from up to down and vice versa but in a different axis.

Pauli X gate electrons go from spin-up to spin-down position by rotating along the x axis; on the contrary, in the case of Pauli Y gate, electrons rotate from spin-up to spin-down position along the y axis in Bloch sphere. Because of the y axis rotation, the imaginary part arrives in the transition matrix and so in the output.

Similar to the Pauli X gate, if a spin-up electron or 0 state electron passes through the Pauli Y gate, then the output from the Pauli Y gate is described in Equation 7.6. We are using the orthogonal vectors described in Equations 7.1 and 7.2 for low- and high-energy state inputs.

For low-energy state input or spin-up electron, the output comes as:

$$\begin{bmatrix} 0 & -i \\ i & 0 \end{bmatrix}\begin{bmatrix} 1 \\ 0 \end{bmatrix} = \begin{bmatrix} 0 \\ i \end{bmatrix} = |\,i> \approx i\,|\,1> \tag{7.6}$$

From Equation 7.6, we can state that the output from the Pauli Y gate is a spin-down electron or a high-energy state with an imaginary number, i, multiplied with it.

Similarly, for a spin-down electron or a low-energy state electron, the output is shown in next:

$$\begin{bmatrix} 0 & -i \\ i & 0 \end{bmatrix}\begin{bmatrix} 0 \\ 1 \end{bmatrix} = \begin{bmatrix} -i \\ 0 \end{bmatrix} = -i\,|\,0> \tag{7.7}$$

From all these discussions, we can compare Pauli X gate and Y gate by the phase angle, and this happens due to the axis of rotation. In the case of Pauli X gate, there is no phase angle associated with the rotated electron, but in the case of Pauli Y gate, there is $90°$ or $\pi/2$ radian phase rotation during spinning.

Quantum circuit ssimulation and probability distribution of Pauli Y gate using IBM virtual quantum computer are shown in Figures 7.4 and 7.5, respectively, using a single qubit [13].

In this context, we discuss about the Pauli Z gate. This gate is much different from the previously discussed Pauli X and Y gates. First, we would take

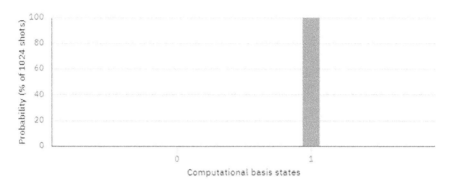

Figure 7.4 Quantum circuit simulation of a single qubit in the Pauli Y gate with |0> qubit as the input.

Figure 7.5 Probability distribution of a single qubit in the Pauli Y gate with |0> qubit as the input.

a look at the transition matrix of the Pauli Z gate and the Dirac representation of it:

$$\begin{bmatrix} 1 & 0 \\ 0 & -1 \end{bmatrix} \approx |0><0| - |1><1|$$

Due to 180° or π radian phase, shifting this gate is referred to as the phase shift gate, sometimes called the phase flipped gate. Using this gate, an electron rotates along the z axis. With |0> as the input, this gate output remains unchanged, and with |1> input, the output becomes –|1>.

Using |0> and |1> as the input in the Pauli Z gate is shown next:

$$|0> \begin{bmatrix} 1 & 0 \\ 0 & -1 \end{bmatrix} = \begin{bmatrix} 1 \\ 0 \end{bmatrix} = |0> \tag{7.8}$$

$$|1> \begin{bmatrix} 1 & 0 \\ 0 & -1 \end{bmatrix} = \begin{bmatrix} 0 \\ -1 \end{bmatrix} = |1> \tag{7.9}$$

Circuit simulation using IBM virtual quantum computer for the Pauli Z gate is shown in Figure 7.6 with <0| input and Figure 7.7 with <1| input:

In Figure 7.7, as |1> is not available in IBM circuit simulation, we use a Pauli X gate or NOT gate to the |0> input after the Pauli Z gate is applied.

Figure 7.6 Quantum circuit simulation of a single qubit in the Pauli Z gate with |0> qubit as the input.

Figure 7.7 Quantum circuit simulation of a single qubit in the Pauli Z gate with |1> qubit as the input.

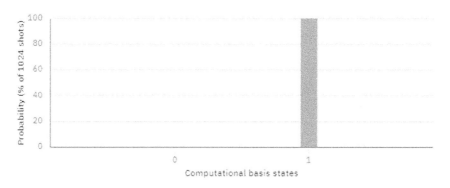

Figure 7.8 Probability distribution of a single qubit in the Pauli Z gate with |1> qubit as the input.

As discussed earlier, with |1> as the input, the resultant output is |1> with π radian phase shift or –|1> (shown in probability distribution and Q-Sphere in Figures 7.8 and 7.9, respectively).

Before we proceed to the next context, we need to know about the concept of eigenvectors and eigenvalue. We know that multiplication of a vector by a matrix results in the vector always, but in some special cases, multiplication of a vector by a matrix leads to the multiplication of a scalar with a vector:

$$A \underset{v}{\rightarrow} = \lambda \underset{v}{\rightarrow}$$

where A is the transformation matrix, λ is the eigenvalue and $\underset{v}{\rightarrow}$ is the eigenvector from the equation.

From Equation 7.7, the output from the Pauli Z gate with |0> input includes the eigenvalue of 1 and the eigen vector of |0>. On the contrary,

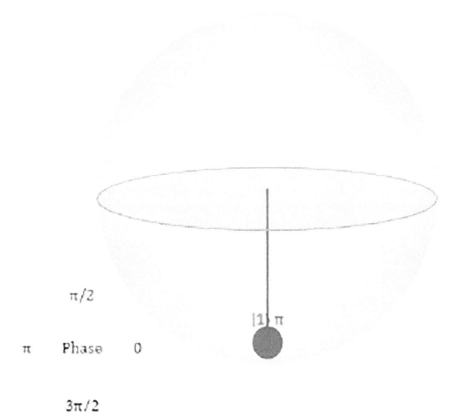

Figure 7.9 Q-Sphere for phase observation of a single qubit in the Pauli Z gate with |1> qubit as the input.

from Equation 7.8, the eigen value is –1, and the eigen vector is |1>. Similarly, for the Pauli X gate, the eigenvectors |+> and |–> are briefed next:

$$|+> = \frac{1}{\sqrt{2}}(|0> + |1>) = \frac{1}{\sqrt{2}}\begin{bmatrix} 1 \\ 1 \end{bmatrix} \tag{7.10}$$

$$|-> = \frac{1}{\sqrt{2}}(|0> - |1>) = \frac{1}{\sqrt{2}}\begin{bmatrix} 1 \\ -1 \end{bmatrix} \tag{7.11}$$

7.4.2 Hadamard (H) Gate

In Pauli X, Y and Z gates, a single qubit is used as the input, and the output is either having reverse spin of the input or the same with some phase difference, so the superposition theory is not likely applied for getting output from those gates. Hadamard gate or H gate has the superposition applied to it for getting the output. In this context, we make a brief discussion about the H gate [14].

The H gate transition matrix is shown next:

$$\frac{1}{\sqrt{2}}\begin{bmatrix} 1 & 1 \\ 1 & -1 \end{bmatrix} \tag{7.12}$$

H gate transforms the qubit states between the x and z axes, so the output can be either high state or low state. The output probability with |0> or low state will be 50% in both low and high states without any phase difference; on the contrary, with |1> input, the output probability leads to the same output as |0> but with 180° or π radian phase difference with |1> or high state. The perfect output will be always 50% probable to each state, but in the real world application, this result may differ [15]. Like all Pauli gates, the H gate also takes one input and gives us one output but with a superposition attached.

Applying |0> and |1> to the H gate leads to:

$$\begin{bmatrix} 1 \\ 0 \end{bmatrix} \frac{1}{\sqrt{2}} \begin{bmatrix} 1 & 1 \\ 1 & -1 \end{bmatrix} = \frac{1}{\sqrt{2}} \begin{bmatrix} 1 \\ 1 \end{bmatrix} \tag{7.13}$$

$$\begin{bmatrix} 0 \\ 1 \end{bmatrix} \frac{1}{\sqrt{2}} \begin{bmatrix} 1 & 1 \\ 1 & -1 \end{bmatrix} = \frac{1}{\sqrt{2}} \begin{bmatrix} 1 \\ -1 \end{bmatrix} \tag{7.14}$$

The output from the H gate with |0> is $\frac{1}{\sqrt{2}}\begin{bmatrix} 1 \\ 1 \end{bmatrix}$, which can also be described as |+>, and the output from |1> input is |->, as discussed in Equations 7.10 and 7.11.

Now, we take a look into the circuit simulation in Figures 7.10–7.15 to have a clear view on the superposition and the working principle of H gate [16].

In Figure 7.10, we show the circuit simulation of the H gate with |0> input, and in Figure 7.11, we show the quantum sphere with that input. In the quantum sphere, it is clearly observable that the output is both |0> and |1> at the same time; this visualization proves by the probability distribution

Figure 7.10 Quantum circuit simulation of Hadamard gate with |0> as the input qubit.

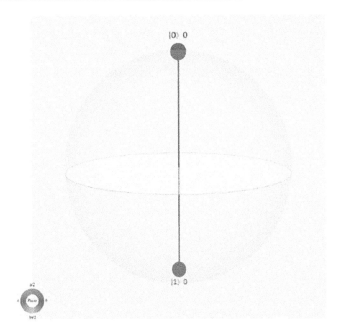

Figure 7.11 Quantum Sphere of Hadamard gate with |0> as the input qubit.

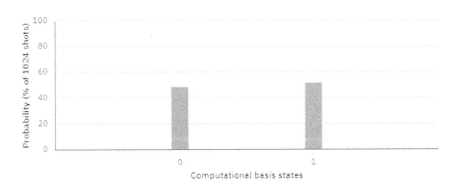

Figure 7.12 Probability distribution of Hadamard gate with |0> as the input qubit.

shown in Figure 7.12, where the probability index versus states has equal 50% probability to be the output in both |0> and |1> states at the same time. Now we look into the simulation, Q-Sphere and probability distribution for |1> state as the input in the H gate.

Circuit simulation for |1> input is shown in Figure 7.13, and the Q-Sphere is shown in Figure 7.14, which supports the theory we discussed earlier, regarding the output of the H gate with |1> input into it. The red-colored line represents the |1> or high-state output with a phase shift of $180°$ or π

Figure 7.13 Quantum circuit simulation of Hadamard gate with |1> as the input qubit.

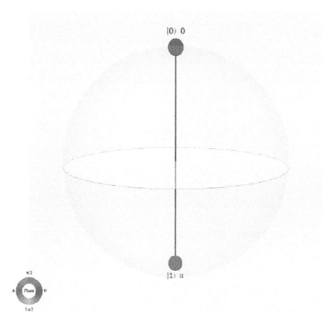

Figure 7.14 Quantum sphere of Hadamard Gate with |1> as the input qubit.

radian. The blue line indicates the output as |0> without any phase shift taking place. The probability distribution from Figure 7.15 is similar to the |0> input with an equal probability of getting |0> and |1> states from the gate output.

7.4.3 R φ Gate or RZ Gate

Rφ gate or RZ gate is quite different from the gates we previously discussed here. Output from this gate can be controlled by a parameter named φ, which must be in the real value range. The transition matrix of RZ gate is as shown next:

$$R\phi = \begin{bmatrix} e^{-i\phi/2} & 0 \\ 0 & e^{i\phi/2} \end{bmatrix}$$

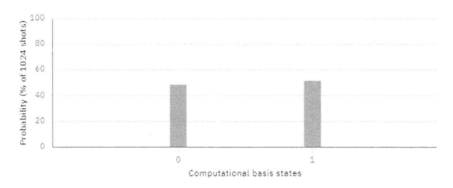

Figure 7.15 Probability distribution of Hadamard Gate with |1> as the input qubit.

where ϕ is a real number. This parameter determines how much the rotation will take place or the spin rotation angle of the input electron.

As we know, $e^{i\phi/2}$ and $e^{-i\phi/2}$ can be defined as:

$$e^{i\phi/2} = \cos\phi/2 + i\,\sin\phi/2 \qquad (7.15)$$

$$e^{-i\phi/2} = \cos\phi/2 - i\,\sin\phi/2 \qquad (7.16)$$

Now, we discuss about the spin rotation and phase shift of the output electron for different values of ϕ.

  a. For $\phi = 0$:

For rotation angle 0, the value of $e^{-i\phi/2}$ and $e^{i\phi/2}$ will be 1 (from Equations 7.15 and 7.16) as the value of cos(0) is 1 and sin(0) is 0. The transition matrix of the RZ gate becomes:

$$\begin{bmatrix} 1 & 0 \\ 0 & 1 \end{bmatrix}$$

If we apply |0> as the input to the RZ gate in this condition, the result becomes:

$$\begin{bmatrix} 1 \\ 0 \end{bmatrix}\begin{bmatrix} 1 & 0 \\ 0 & 1 \end{bmatrix} = \begin{bmatrix} 1 \\ 0 \end{bmatrix} = |0>$$

So we get |0> as the result.

Similarly, if we apply |1> to the transition matrix, we get,

$$\begin{bmatrix} 0 \\ 1 \end{bmatrix}\begin{bmatrix} 1 & 0 \\ 0 & 1 \end{bmatrix} = \begin{bmatrix} 0 \\ 1 \end{bmatrix} = |1>$$

So, apparently, with $\phi = 0$, RZ gate is not converting the state of the input qubit with respect to magnitude, but there might be a phase difference of $360°$ or 2Å radian to the output from its gate [16]. We look at the output from circuit simulation from the IBM virtual quantum computer in the next context, but prior to that, we will discuss about the unity matrix and its use over the quantum gates.

We all know about the unity matrix or unitary matrix, and it looks like:

$$\begin{bmatrix} 1 & 0 \\ 0 & 1 \end{bmatrix}$$

This matrix is obtained from the RZ gate with $\phi = 0$ as the unitary matrix. Comparing the output with the unitary matrix, we can also determine the change in the state and the magnitude as well as the phase. See Figure 7.16.

RZ gate output with condition $\phi = 0$ for |1> is shown in Figure 7.17. Comparing with the unitary matrix described earlier, we may conclude that there will be a 2π radian phase difference present with the output. As the output probability is completely high state, we can recall the parameter from Equation 7.2, the probability p has the value of 1, and now we have one parameter ϕ; we consider the phase shift of the output qubit as θ here to avoid the conflict. Now, if we rewrite Equation 7.2 with these parameters, we get:

$$| \psi> = e^{j\theta} |1> \tag{7.17}$$

With separating $e^{j\cdot}$ in imaginary parts, we get: $\cos\theta + j\sin\theta$; rewriting Equation 7.17 with replacing the exponential part and the matrix format of |1>, the output state of the qubit became:

$$| \psi >= (\cos\theta + j\sin\theta)\begin{bmatrix} 0 \\ 1 \end{bmatrix} \tag{7.18}$$

Figure 7.16 Quantum circuit simulation of RZ gate for $\phi = 0$ with |0> as the input qubit.

Figure 7.17 Quantum circuit simulation of RZ gate for $\phi = 0$ with ||1> as the input qubit.

As the output does not contain any imaginary part, the value of sinθ will be 0, and on the contrary, cosθ must be equal to 1. With these conditions, the possible values of θ will be either zero or 2π radian. Apparently, from the circuit simulation output, we get the 2π radian phase shift to the output qubit state, which supports the calculation we have done here.

b. For $\phi = \pi/4$:

For this condition, the transition matrix of the RZ gate becomes:

$$\begin{bmatrix} 0.9238 - i0.3826 & 0 \\ 0 & 0.9238 + i0.3826 \end{bmatrix} \tag{7.19}$$

With the low input state, the output we get from the RZ gate with $\phi = \pi/4$:

$$\begin{bmatrix} 1 \\ 0 \end{bmatrix}\begin{bmatrix} 0.9238 - i0.3826 & 0 \\ 0 & 0.9238 + i0.3826 \end{bmatrix} = \begin{bmatrix} 0.9238 - i0.3826 \\ 0 \end{bmatrix}$$

This expression is equivalent to:

$$(0.9238 - i0.3826)\begin{bmatrix} 1 \\ 0 \end{bmatrix} = (0.9238 - i0.3826)\,|\,0 > \tag{7.20}$$

We can conclude that the output qubit state is actually zero state or low-energy state associated with some magnitude or amplitude and phase difference. We will also verify calculations with the circuit simulation of RZ gate with $\phi = \pi/4$ radian.

Amplitude or magnitude of the output qubit state will be $\sqrt{(0.9238)^2 + (0.3826)^2} = \sqrt{(0.85340644) + (0.14638276)} = \sqrt{0.9997892} \approx \sqrt{1} = 1$

Phase difference of the output qubit state will be $\tan^{-1}\left(-\dfrac{0.3826}{0.9238}\right) =$ –22.49732717; if we depict the phase angle in the real and imaginary axes, we get:

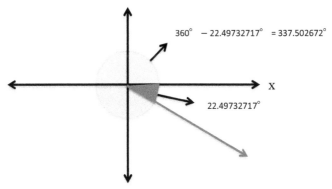

From the discussion from the top, we get that the calculated phase angle 22.49732717° is similar to the angle 337.502672°.

Now, we simulate the RZ gate circuit with $\phi = \frac{\pi}{4}$ and check the result with our calculated values of amplitude and phase angle.

The output state from Figure 7.18 matched perfectly with Equation 7.20. Also the state vector and quantum sphere support the 100% probability of being 0 in the output state.

From Figure 7.19, we get that the phase angle from the output qubit state is $\frac{15\pi}{8}$, which is equal to the angle we calculated earlier, 337.502672°.

With the high input state, the output we get from the RZ gate with $\phi = \frac{\pi}{4}$:

$$\begin{bmatrix} 0 \\ 1 \end{bmatrix} \begin{bmatrix} 0.9238 - i0.3826 & 0 \\ 0 & 0.9238 + i0.3826 \end{bmatrix} = \begin{bmatrix} 0 + i0 \\ 0.9238 + i0.3826 \end{bmatrix}$$

Figure 7.18 Quantum circuit simulation of RZ gate for $\phi = \frac{\pi}{4}$ with |0> as the input qubit.

Figure 7.19 Quantum sphere of RZ gate for $\phi = \frac{\pi}{4}$ with |0> input.

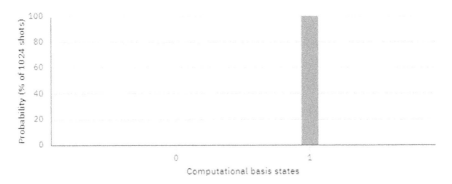

Figure 7.20 Probability distribution of RZ gate for $\phi = \frac{\pi}{4}$ with $|1>$ as the input qubit.

Like we calculate the magnitude of RZ gate with $\phi = \frac{\pi}{4}$ with low input, similarly for high input state, the magnitude is 1, due to the same values, as the negative sign with an imaginary number does not have any effect on the magnitude.

With real and imaginary values, 0.9238 and 0.3826, the phase angle will be different in this case for high input. The phase angle of the output qubit is $\tan^{-1}\left(\dfrac{0.3826}{0.9238}\right) = 22.49732717$. Now let's take a look at the circuit simulation for the RZ gate with $\phi = \frac{\pi}{4}$ with high input.

In Figure 7.20, we can clearly see the output state values that match our calculation and 100% possibility of output quantum state for being in $|1>$. On the contrary, on the right hand side, the output phase angle $\frac{\pi}{8}$ or 22.49732717 also matched with phase angle calculation. In the next context, we discuss the RZ gate with $\phi = \frac{\pi}{2}$.

c. For $\phi = \frac{\pi}{2}$:

With $\phi = \frac{\pi}{2}$, the RZ gate transition matrix becomes:

$$\begin{bmatrix} 0.707106 - i0.707106 & 0 \\ 0 & 0.707106 + i0.707106 \end{bmatrix}$$

Similar operations, like in the previous cases for low and high inputs, with the output qubit state are described in Equations 3.20 and 3.21, respectively:

$$\begin{bmatrix} 1 \\ 0 \end{bmatrix}\begin{bmatrix} 0.707106 - i0.707106 & 0 \\ 0 & 0.707106 + i0.707106 \end{bmatrix}$$
$$= (0.707106 - i0.707106)\begin{bmatrix} 1 \\ 0 \end{bmatrix} \qquad (7.21)$$

$$\begin{bmatrix} 0 \\ 1 \end{bmatrix} \begin{bmatrix} 0.707106 - i0.707106 & 0 \\ 0 & 0.707106 + i0.707106 \end{bmatrix}$$

$$= (0.707106 + i0.707106) \begin{bmatrix} 0 \\ 1 \end{bmatrix} \qquad (7.22)$$

In these scenarios, a constant $(0.707106 - i0.707106)$ and $(0.707106 + i0.70710)$ is multiplied with the |0> and |1> states, respectively, in the output. If we calculate the magnitude of the outputs likewise in the previous cases, it will be 1. In the case of phase difference with the low input state, the phase difference will be $tan^{-1}\left(-\dfrac{0.707106}{0.707106}\right) = -45°$ or $315°$ and will be $\dfrac{7\pi}{4}$ radian. Similarly, for the high input case, the phase difference will be $tan^{-1}\left(\dfrac{0.707106}{0.707106}\right) = 45°$ and will be $\dfrac{\pi}{4}$ radian. The simulation output from the IBM virtual quantum computer is shown in Figures 7.21–7.24 for low and high input states.

We can conclude by observing these simulations that our calculated values matched perfectly with the simulated values.

d. For $\phi = \dfrac{3\pi}{4}$:

Now, we discuss for the case where the value of ϕ is $\dfrac{3\pi}{4}$. In that case, the transition matrix becomes:

$$\begin{bmatrix} 0.3826834 - i0.92387953 & 0 \\ 0 & 0.3826834 + i0.92387953 \end{bmatrix}$$

Figure 7.21 Quantum circuit simulation of RZ gate for $\phi = \dfrac{\lambda}{2}$ with |0> as the input qubit.

Figure 7.22 Quantum circuit simulation of RZ gate for $\phi = \dfrac{\lambda}{2}$ with ||> as the input qubit.

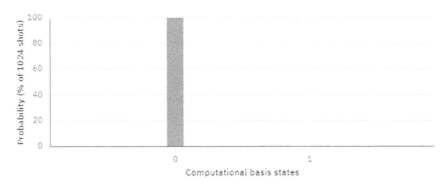

Figure 7.23 Probability distribution of RZ gate for $\phi = \frac{\pi}{2}$ with |0> as the input qubit.

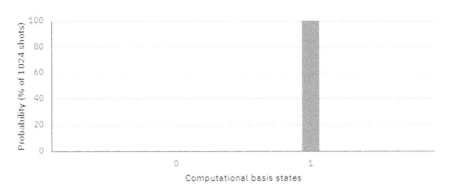

Figure 7.24 Probability distribution of RZ gate for $\phi = \frac{\pi}{2}$ with |1> as the input qubit.

With applying |0> and |1> as the input state, the output state with respect to this transition matrix is calculated in Equations 3.22 and 3.23, respectively:

$$\begin{bmatrix} 1 \\ 0 \end{bmatrix} \begin{bmatrix} 0.3826834 - i0.92387953 & 0 \\ 0 & 0.3826834 + i0.92387953 \end{bmatrix}$$
$$= (0.3826834 - i0.92387953) \begin{bmatrix} 1 \\ 0 \end{bmatrix} \tag{3.22}$$

$$\begin{bmatrix} 0 \\ 1 \end{bmatrix} \begin{bmatrix} 0.3826834 - i0.92387953 & 0 \\ 0 & 0.3826834 + i0.92387953 \end{bmatrix}$$
$$= (0.3826834 + i0.92387953) \begin{bmatrix} 0 \\ 1 \end{bmatrix} \tag{3.23}$$

In this case, the magnitude is also 1 as in the previous input cases. And the phase difference is $tan^{-1}\left(-\dfrac{0.92387953}{0.3826834}\right) = -67.50$ or 292.49 in the low input case; on the contrary, for the high input case, this phase difference is 67.50. Now, we simulate this in the IBM quantum simulator and verify the output with our calculated values. Simulation outputs with probability distributions and Q-Sphere are shown in the Figures 7.25–7.30 for low and high inputs, respectively.

As Figures 7.27 and 7.30 suggest, the output magnitude for both the low and the high inputs is 1, and the phase difference is $\dfrac{13\pi}{8}$ and $\dfrac{3\pi}{8}$ for low and high inputs, respectively. These results match with our calculated values.

With the described options here for the parameter in the RZ gate, we show some possible operations from different aspects. With calculated values, circuit simulation confirms the results. There can be many values starting from 0 to 360 that can be applied. In every case, the magnitude is 1, but the phase difference changes, and as a result, the rotation angle differs.

In the next context, we discuss about other gates: Fredkin gate.

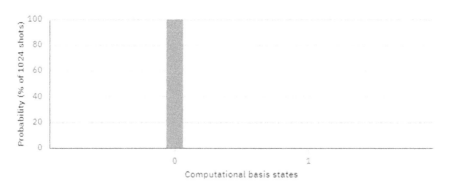

Figure 7.25 Quantum circuit simulation of RZ gate for $\phi = \dfrac{3\pi}{4}$ with |0> as the input qubit.

Figure 7.26 Probability distribution of RZ gate for $\phi = \dfrac{3\pi}{4}$ with |0> as the input qubit.

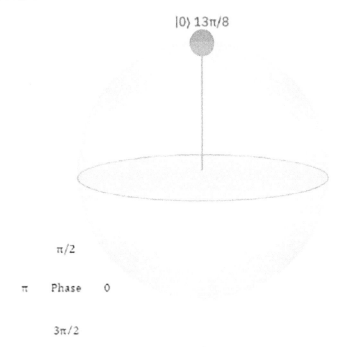

|0⟩ 13π/8

π/2

π   Phase   0

3π/2

Figure 7.27 Q-Sphere of RZ gate for $\phi = \frac{3\pi}{4}$ with |0> as the input qubit.

Figure 7.28 Quantum circuit simulation of RZ gate for $\phi = \frac{3\pi}{4}$ with |1> as the input qubit.

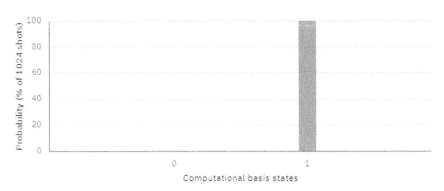

Figure 7.29 Probability distribution of RZ gate for $\phi = \frac{3\pi}{4}$ with |1> as the input qubit.

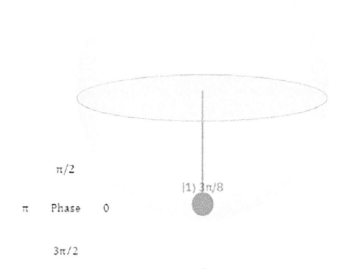

Figure 7.30 Q-Sphere of RZ gate for $\phi = \frac{3\pi}{4}$ with |1> as the input qubit.

7.5 MULTIPLEXER USING QUANTUM BITS

After a brief discussion on different quantum logic gates using the qubits, let's make a combinational logic circuit using the quantum gates. From the logic families of combinational logic circuits, we try to explain the quantum circuit diagram of Multiplexer along with the probability of getting the output states and the truth table comparison [17]. Starting with a 2 × 1 Multiplexer, we dive into the complex circuit simulation of 4 × 1 and 8 × 1 Multiplexer as well. In Section 7.5.1, we discuss about the main gate that is needed to proceed with the current discussion: the Fredkin gate or CSWAP gate.

7.5.1 Fredkin Gate

Fredkin gate, which is also known as CSWAP gate or the conservative logic gate, is a reversible gate introduced by Edward Fredkin. This gate consists of three inputs and three outputs related to the input qubits in a specific manner. This gate accepts three inputs denoted as input control (IC) qubit along with two data inputs with symbols I_0 and I_1. Output qubits are denoted

Table 7.3 Truth Table of Fredkin or CSWAP Gate.

Input control qubit	Input 1	Input 2	Output control qubit	Output 1	Output 2
0	0	0	0	0	0
0	0	1	0	0	1
0	1	0	0	1	0
0	1	1	0	1	1
1	0	0	1	0	0
1	0	1	1	1	0
1	1	0	1	0	1
1	1	1	1	1	1

as output control (OC) qubit along with two data output qubits O_0 and O_1. The control qubit remains unchanged in both the input and the output sections, but the output qubits O_0 and O_1 mapped with the input qubits I_0 and I_1 when the control qubit is in the low state or 0 state [18]. This mapping gets swapped when the control qubit is in 1 state or high state; during the control qubit in 1 state, the output qubits O_0 and O_1 get mapped with the input qubits I_1 and I_0, respectively. Let's check the truth table of this particular gate in the Table 7.3.

As we described in the previous section, the truth table in Table 7.3 describes the similar operation. In the first four cases, the input control qubits are in 0 state; during that time, output 1 and output 2 become the replica of input 1 and input 2. In the last four cases, the input control qubit is 1; in that case, output 1 and 2 got swapped and mapped with input 2 and 1, respectively [19]. Due to the swapping characteristic, this gate is referred to as the CSWAP gate.

7.5.2 Multiplexer

Multiplexing means transmitting a large number of information units over a small number of channels or lines. The multiplexer is a combinational circuit that serves this purpose.

A multiplexer is a combinational logic circuit having 2^n input lines with a single output line. The selection of input to output transfer path is controlled by a set of n select lines. In Figures 7.31–7.33, we show the block diagram of a 2 × 1 MUX, followed by 4 × 1 and 8 × 1 MUX, along with the truth tables (Tables 7.4–7.6).

Let's discuss about the operation principle of the Multiplexer in this context. Starting with the 2 × 1 Multiplexer, this particular design has one select input; based on the value of the select input, one of the two input values is redirected to the output line; as per the truth table of the 2 × 1 multiplexer, when the select input is 0, I_0 is redirected to the output line and I_1 is

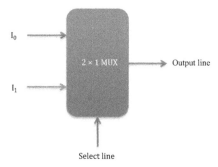

Figure 7.31 Block diagram of 2 × 1 Multiplexer with 2 input lines, 1 select line and 1 output line.

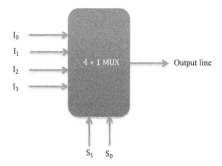

Figure 7.32 Block diagram of 4 × 1 Multiplexer with 4 input lines, 2 select lines and 1 output line.

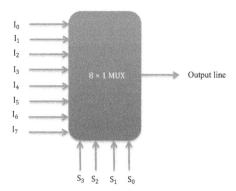

Figure 7.33 Block diagram of 8 × 1 Multiplexer with 8 input lines, 4 select lines and 1 output line.

Table 7.4 Truth Table of 2 × 1 MUX with 2 Input Lines with Select Input and Single Output Lines.

Input 1	Input 2	Select input	Output
I_0	I_1	0	I_0
I_0	I_1	1	I_1

Table 7.5 Truth Table of 4 × 1 MUX with 4 Input Lines with 2 Select Inputs and Single Output Line.

Input 1	Input 2	Input 3	Input 4	Select input 1	Select input 2	Output
I_0	I_1	I_2	I_3	0	0	I_0
I_0	I_1	I_2	I_3	0	1	I_1
I_0	I_1	I_2	I_3	1	0	I_2
I_0	I_1	I_2	I_3	1	1	I_3

Table 7.6 Truth Table of 8 × 1 MUX with 8 Input Lines with 3 Select Inputs and Single Output Line.

Input 1	Input 2	Input 3	Input 4	Input 5	Input 6	Input 7	Input 8	Select input 1	Select input 3	Select input 4	Output
I_0	I_1	I_2	I_3	I_4	I_5	I_6	I_7	0	0	0	I_0
I_0	I_1	I_2	I_3	I_4	I_5	I_6	I_7	0	0	1	I_1
I_0	I_1	I_2	I_3	I_4	I_5	I_6	I_7	0	1	0	I_2
I_0	I_1	I_2	I_3	I_4	I_5	I_6	I_7	0	1	1	I_3
I_0	I_1	I_2	I_3	I_4	I_5	I_6	I_7	1	0	0	I_4
I_0	I_1	I_2	I_3	I_4	I_5	I_6	I_7	1	0	1	I_5
I_0	I_1	I_2	I_3	I_4	I_5	I_6	I_7	1	1	0	I_6
I_0	I_1	I_2	I_3	I_4	I_5	I_6	I_7	1	1	1	I_7

redirected to the output line during the value of the select line 1. In the case of the 4 × 1 Multiplexer, two select inputs are present; when both the select inputs are 0, I_0 is redirected to the output line; when the first select input is 0 and the second one is 1, I_1 is redirected to the output line and so on. This scenario is briefly described in the truth table of the 4 × 1 Multiplexer in the aforementioned context. Similarly, for the 8 × 1 Multiplexer, there are 8 inputs along with 3 select inputs, which is also tabulated.

As we can clearly see, the Multiplexer or MUX has similar characteristics of a switch, the select inputs act as the main parameter in this case and the input and output lines are connected based on this select line.

Now, we will try to continue a brief discussion about the Multiplexer circuit design using quantum bits or qubits. The previously mentioned 2 × 1 Multiplexer can be easily created using the Fredkin gate or the CSWAP gate. If we get a closer look in the truth table of the Fredkin gate or CSWAP gate, if we consider the input control qubit as the select input and other two input qubits as the inputs in the Multiplexer, and the output can be collected from output 1 from the truth table, without any further discussion, let's dive into the quantum circuit simulation of 2 × 1 MUX using qubits.

In the quantum circuit diagram in Figure 7.34 of 2 × 1 Multiplexer, q[0] is considered the control qubit, and q[1] and q[2] are considered two input qubits with the value of low state or |0>. As per the truth table of the Fredkin gate, the control qubit remains unchanged, and both q[1] and q[2] also remain unchanged. The Multiplexer output is collected from q[1], and from the probability distribution (Figure 7.35), we can observe that it remains in the |0> state, and as per the 2 × 1 Multiplexer logic, the value of q[1] is returned in the output line, which satisfies the 2 × 1 Multiplexer logic successfully for both the inputs in the |0> state or the low-energy state.

Let's check the circuit diagram with probability distribution for the |0> and |1> states in the q[1] and q[2] qubits, simultaneously, in the 2 × 1 Multiplexer (see Figures 7.36 and 7.37).

Figure 7.34 Quantum circuit diagram of 2 × 1 Multiplexer for both inputs in |0> state using the Fredkin gate.

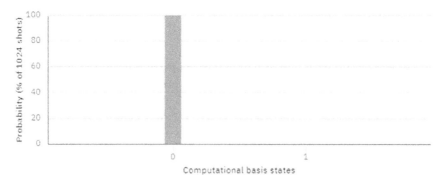

Figure 7.35 Probability distribution of 2 × 1 Multiplexer made in quantum circuit using the Fredkin gate for both the input qubits in |0> state.

Figure 7.36 Quantum circuit diagram of 2 × 1 Multiplexer with |0> and |1> states in the q[1] and q[2] qubits, simultaneously.

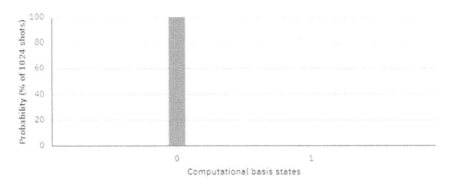

Figure 7.37 Probability distribution of quantum circuit of 2 × 1 Multiplexer with control input in |0> state with |0> and |1> states in the q[1] and q[2] qubits, simultaneously.

As per the discussion, the |0> state fed into the q[1] qubit, and the initial |0> state converted into the |1> state using the Pauli X gate and fed into the q[2] qubit, and the control qubit q[0] stays at the |0> state like earlier. As per the logic of the 2 × 1 Multiplexer, the output should be the value of q[1], which is actually the |0> state in this case. The probability distribution shows that the maximum probability of getting the |0> state from this circuit is 100%, which supports the multiplexer logic for this case.

Let's discuss the quantum circuit diagram and the probability distribution with |1> and |0> as input in q[1] and q[2] qubits, when the control qubit q[0] has the |0> state (see Figures 7.38 and 7.39).

In this case, when the inputs in q[1] and q[2] qubits are |1> and |0> states, respectively, with control qubit q[0] having the input of the |0> state, the output probability from this circuit is the value of q[1] qubit, which is the |1> state. This output with this circuit configuration actually is the perfect match with the logic of 2 × 1 Multiplexer.

In this context, we will change the control qubit with the value of |1> state, with q[1] having the value of |0> state and q[2] having the value of |1> state. As per the 2 × 1 Multiplexer logic, the value of q[2] qubit should be present in the output line. Let's check the circuit diagram and probability distribution with these inputs in Figure 7.40.

Figure 7.38 Quantum circuit diagram of 2 × 1 MUX with |1> and |0> states as input in q[1] and q[2] qubits when the control qubit has the |0> state.

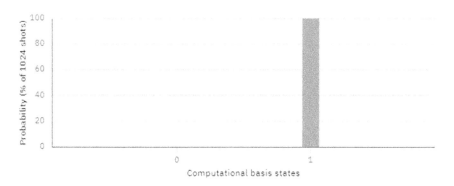

Figure 7.39 Probability distribution of 2 × 1 MUX with |1> and |0> states as input in q[1] and q[2] qubits when the control qubit has the |0> state.

Figure 7.40 Quantum circuit diagram of 2 × 1 MUX with |0> and |1> states as input in q[1] and q[2] qubits when the control qubit has the |1> state.

From Figure 7.41, the probability distribution shows the maximum probability of getting the |1> state from the output line, which is the value of the q[2] qubit. It supports the discussion in this section.

From all the discussions regarding the 2 × 1 MUX implementation, using quantum gates with qubits as the inputs, we can say that the circuit simulation we present worked perfectly for implementing the 2 × 1 Multiplexer logic. When the control qubit is low, it selects the first input qubit (q[1]); also during high state input in the control qubit, this circuit selects the second input qubit (q[2]).

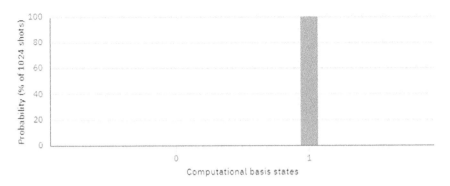

Figure 7.41 Probability distribution of 2 × 1 MUX with |0> and |1> states as input in q[1] and q[2] qubits when the control qubit has the |1> state.

7.6 CONCLUSION

This chapter explains the chronological development of combinational circuits with various probabilistic outcomes. For the sake of circuit design, specifics of the quantum logic gates that will be used in a future quantum computer are described. The architecture and ensuing construction of fundamental quantum circuits, as well as quantum computing techniques, are also described. The chapter also shows how the process varies from the traditional methods. The effort may be focused on the hardware implementation of quantum information processing in the future.

REFERENCES

[1] H. Bhatt, S. Gautam, "Quantum computing: A new era of computer science", IEEE 6th International Conference on Computing for Sustainable Global Development, 2019, New Delhi, India
[2] Z. Meng, "Review of quantum computing", IEEE 13th International Conference on Intelligent Computation Technology and Automation, 2020, Xi'an, China
[3] Y. Ding, D. Llewellyn, I. I. Faruque, D. Bacco, K. Rottwitt, M. G. Thompson, J. Wang, L. K. Oxenlowe, "Quantum entanglement and teleportation based on silicon photonics", IEEE 22nd International Conference on Transparent Optical Networks, 2020, Bari, Italy
[4] J. Yu, W. Xu, "Calculation of quantum entanglement", IEEE 10th International Symposium on Distributed Computing and Applications to Business, Engineering and Science, 2011, Wuxi, China
[5] R. Rietsche, C. Dremel, S. Bosch, L. Steinacker, M. Meckel, J.-M. Leimeister, "Quantum computing", Electronic Markets, vol. 32, pp. 2525–2536, 2022
[6] N. Tarek, B. A. Hafid, "The quantum computer and the security of information systems", IEEE International Conference on Recent Advances in Mathematics and Informatics, 2021, Tebessa, AL

[7] M. Iqbal, L. Velasco, M. Ruiz, A. Napoli, J. Pedro, N. Costa, "Quantum bit retransmission using universal quantum copying machine", IEEE International Conference on Optical Network Design and Modelling, 2022, Warsaw, Poland

[8] H. E. Brandt, "Qubit devices and the issue of quantum decoherence", Progress in Quantum Electronics, vol. 22, no. 5–6, pp. 257–370, 1999

[9] S. Shrapnel, F. Costa, G. Milburn, "Updating the Born rule", New Journal of Physics, vol. 20, p. 053010, 2018

[10] D. McMahon, "Quantum gates and circuits", in *Quantum computing explained*, Wiley-IEEE Press, pp. 173–196, 2008, Hoboken, NJ

[11] M. N. Sarfaraj, S. Mukhopadhyay, "All-optical scheme for implementation of tri-state Pauli-X, Y and Z quantum gates using phase encoding", Optoelectronics Letters, vol. 17, pp. 746–750, 2021

[12] P. De, S. Ranwa, S. Mukhopadhyay, "Intensity and phase encoding for realization of integrated Pauli X, Y and Z gates using 2D photonic crystal", Optics & Laser Technology, vol. 152, p. 108141a, 2022

[13] S. Dey, S. Mukhopadhyay, "Implementation of all-optical Pauli-Y gate by the integrated phase and polarisation encoding", IET Optoelectronics, vol. 12, no. 4, pp. 176–179, 2018

[14] N. Abdessaied, M. Soeken, R. Drechsler, "Quantum circuit optimization by Hadamard gate reduction", International Conference on Reversible Computation, vol. 8507, ch. 12, pp. 149–162, 2014

[15] A. Tipsmark, R. Dong, A. Laghout, M. Jezek, U. L. Andersen, "Experimental demonstration of a Hadamard gate for coherent state qubits", IEEE Conference on Lasers and Electro-Optics Europe and 12th European Quantum Electronics Conference, 2011, Munich, Germany

[16] A. Nabeyama, H. Yashima, "All-optical switchable logic gate using a single QD-SOA for RZ-BPSK signal inputs", Optical and Quantum Electronics, vol. 53, no. 244, 2021

[17] Z. Shan, Y. Zhu, B. Zhao, "A high-performance compilation strategy for multiplexing quantum control architecture", Scientific Reports, vol. 12, no. 7132, 2022

[18] B. Safaiezadeh, L. Kettunen, M. Haghparast, "Novel high-performance QCA Fredkin gate and designing scalable QCA binary to gray and vice versa", The Journal of Supercomputing, vol. 79, pp. 7037–7060, 2023

[19] S. Dey, P. De, S. Mukhopadhyay, "An all-optical implementation of Fredkin gate using Kerr effect", Optoelectronics Letters, vol. 15, pp. 317–320, 2019

Chapter 8

Recent Trends and Challenges in Quantum Computing Based on Artificial Intelligence

Krishnanjan Mukherjee, Ratneswar Ghosh, and Soumen Santra

8.1 INTRODUCTION

Before understanding what quantum computing is, it should be well known what quantum theory is. Quantum theory or quantum physics is a modern-day theoretical concept of physics that deals with the fact that both at the atomic and at the subatomic levels, matter and energy exist. This theory was founded by Neils Bohr and Max Plank (also called the founding fathers of quantum theory). Albert Einstein was also considered to be the creator of quantum theory because his work was proposed in his Theory of Photoelectric Effect. As a result, he won the Nobel Prize in 1921. This theory states that radiant energy is divided into finite small pockets of energy known as quanta. These pockets of energy are used in a number of ways such as in energy transfer at the atomic level. It should be noted that the uniqueness of Einstein's work was that he represented light energy in the form of quanta. The entire quantum theory is applicable at the atomicity levels [1–3].

The theory states that through an electron, there are certain energy states that are permissible and are therefore found to be quantized. Moreover, it also states that in a particular system, more than one electron cannot occupy the same energy level. This theory has a number of applications, one of which is that it helps to understand the properties and behaviours of matter and energy at the most fundamental level. Quantum theory has a number of applications in our day-to-day life including quantum chemistry, quantum optics, superconducting magnets, and so on [4, 5].

It has been observed in the realm of computing that classical computers could solve most of the day-to-day problems. But there are certain problems that require a different algorithm to find the solution. David Deutsch, an Oxford physicist, brought into light the theory of quantum computing to prove the existence of a parallel universe [6–8]. Quantum computing is an area under computer science that focuses on the development of technology that works on the principle of quantum theory; it considers the presence of bits (like quanta) to carry out a particular task more efficiently [7–9].

In quantum computation, there are a number of models that are being widely used and are referred to as quantum circuits. One of the quantum

DOI: 10.1201/9781003373117-8

circuits is Adiabatic Quantum Computation (AQC), which is quite like quantum annealing, but it works on the principle of Adiabatic Theorem. Quantum Turing Machine, referred to as the universal quantum computer, is an abstract machine used to draw the significance of quantum computers, and quantum annealing is a type of an optimization technique in which the global minimum is obtained from a particular objective function with respect to a given candidate state. The foundation of these models is that these are based on quantum bits, or "qubits," which are like bits in classical computers [10–12].

AQC is a proposed solution to the problem of energy relaxation. Interference from the outside world cannot cause the quantum system to move to a lower state because it is in the ground state [10–13].

As mentioned earlier, the basis of quantum computing is qubits. A qubit (analogous to quanta) is the general information part of it that contains two energy states: 1 and 0. According to a study by Dan C. Marinescu and Gabriel M. Marinescu in *Classical and Quantum Information* (2012), a q-circuit is composed of a collection of QLogic gates that are interconnected by quantum wires. Studies revealed that the CNOT gate or the Feynman gate is a type of logic gate that is used to develop devices that work on the principles of quantum theory [8–13]. With the help of CNOT gate, along with qubits, any circuit can be simulated to a particular accuracy. This gate was developed by Richard Feynman in 1986, after whom it was named so.

8.1.1 Literature Survey

Recently in a research paper, published by a group of scientists, it has been published that the major aim of quantum computing is to blend fundamental quantum information with its commercial application, majorly in the industrial and finance sectors [1–5]. It has been found out that in today's world, ecologists have the tendency to exploit the application of quanta in ecological studies where the classical statistical theories have certain flaws that can be compensated using quantum computing. Ecologists have a long-standing tradition of accepting new technologies; most lately, deep literacy, machine learning, and artificial intelligence are the arising areas that are being used to enhance our understanding of ecological systems [6–9].

8.1.2 Historical Development of Quantum Computing

In 2020, a paper was published where a detailed analysis of quantum computing application was demonstrated, stating the fact that cryptography application was used in 5G [10, 12–14]. The paper stated that a series of simple, phased, and recommended changes are required to guarantee that the security of 5G network is not exploited even with the introduction of large-scale quantum computation. This paper also proposed a *5G security system design* where a multi-phase approach is described for the upgradation of security and privacy in a post-quantum system [12–15].

In today's world, the major flaw of classical cryptography is that it does not have the ability to solve difficult mathematical problems in a short period of time. The introduction of quantum computing in cryptography makes things easier because it considers quantum physics laws to disapprove many asymmetric algorithms used in the classical method [11].

A series of technology and protocols were proposed by a group of studies in a research paper where they have discussed the application of Quantum Key Distribution (QKD) systems along with its security implication. In this paper, it was mentioned that the main building block of quantum encryption system is Quantum Random Number Generators (QRNGs) [13–17].

Studies reveal that in cryptography, constraints such as sequence ciphers, block ciphers, and hash functions are used. In cryptography, the designing of Boolean function works as the building block. It has been noticed that during the designing process, due to the super exponential size $\theta(2^{2n})$ of the space, searching the variable n is computationally hard. Recently, in a paper, with the help of ground state of Ising Hamiltonian, a codification of the constraints was proposed, which proved to provide a speed up in quantum [14–16, 18].

Researchers reveal that quantum computing and deep learning are the hot topics in recent days. In 2020, quantum computing–based deep learning methods were introduced to facilitate fault detection. The fault diagnosis model works on the principle of generative training assisted by quantum computation. To justify its application and output efficiency, the model was implemented in a *Continuous Stirred Tank Reactor (CSTR)* and Tennessee Eastman (TE) processes for the purpose of monitoring. The statistical figure, as stated and proved in the paper, says that the proposed model inhibits a superior fault diagnosis performance with a success rate of 79% and 99.4%, respectively, for the CSTR and TE processes [14–19].

In the field of Big Data Analytics, Artificial Neural Network (ANN) implementation is one of the most efficient ways of solving problems. In ANN, the complexity and nonlinear feature of the input can be generalized and modified. But due to the availability of enormous data coming from a number of unknown sources, supercomputers fail to train ANN due to its enormous size and dimension. Moreover, a large number of complex algorithms are to be studied to analyse different patterns. Therefore, in quantum computing, data is stored in the form of *qubits* [17–20]. Qubits in quantum computers are implemented to detect various patterns easily for analysis, which was little cumbersome for classical computers. In 2020, a research paper was published wherein a simulation program was run in order to modify the application of ANN in Big Data Analytics. The simulation output, hence proved, that the QC approach of ANN has a higher efficiency than the classical one [15–18, 21].

The architecture of software-intensive systems allows architects to use architecture-centric processes, practices, and description languages to model, design, and advance quantum computing software (quantum software for short) at a higher level of abstraction [20].

We performed a systematic literature review (SLR) to identify (i) architectural processes, (ii) modelling notation, (iii) architectural design patterns, (iv) tool support, and (iv) challenge factors for the quantum software architecture [21–24].

The SLR results show that quantum software represents a new genre of software-intensive systems. However, existing processes and notations can be applied to derive architectural activities and develop modelling languages for quantum software.

The results of this SLR will help researchers and practitioners develop new hypotheses to test, derive reference architectures, and use architecture-centric principles and practices to develop new and next-generation quantum software.

8.2 ESSENTIAL HARDWARE COMPONENTS OF A QUANTUM COMPUTER

8.2.1 Data Plane of Quantum

A QC's "heart" is its quantum data plane. It contains the physical qubits as well as the structures required to keep them in place. It must also include any support circuitry required to measure the state of the qubits and conduct gate operations on the physical qubits in a gate-based system or to control the Hamiltonian in an analogue computer. Control signals sent to the chosen qubit(s) set the Hamiltonian it perceives, which controls the gate operation of a digital quantum computer. Because some qubit operations require two qubits in gate-based systems, the quantum data plane must provide a programmable "wiring" network that allows two or more qubits to interact. Analogue systems frequently require greater communication between qubits, which this layer must accommodate. Because high qubit fidelity necessitates strong isolation from the environment, connectivity is limited—it may not be possible for every qubit to interact directly with every other qubit—so the computation must be mapped to the specific architectural constraints of this layer. Because of these limits, operation integrity and connection are critical characteristics of the quantum data layer. Unlike in a traditional computer, where both the control plane and the data plane components use the same silicon technology and are integrated on the same device, control of the quantum data plane requires a different technology than that of the qubits and is performed externally by a separate control and measurement layer (which is described next). The analogue control information for the qubits must be supplied to the proper qubit (or qubits). This control information is delivered electrically in some systems via wires, making these cables a part of the quantum data plane; in others, it is conveyed via optical or microwave radiation. Transmission must be implemented with high specificity, such that it affects only the intended qubit(s) without interfering with the other qubits in the system. As the number of qubits increases, this becomes more challenging; the number of qubits in a single module is thus another essential element of a quantum data layer [5–9].

8.2.2 Parameters of Plane of Control and Measurement

The control and measurement plane translates the digital signals from the control processor, which indicate which quantum operations are to be conducted, to the analogue control signals required to perform the operations on the qubits in the quantum data plane. It also translates the analogue output of data plane qubit measurements to conventional binary data that the control processor can handle. Because of the analogue nature of quantum gates, the production and transmission of control signals are difficult; tiny errors in control signals, or anomalies in the physical architecture of the qubit, will alter the results of operations. As the machine runs, the errors associated with each gate operation add up. Any flaw in the isolation of these signals (so-called signal crosstalk) will result in the appearance of minor control signals for qubits that should not normally be addressed during an operation, resulting in slight mistakes in their qubit state [5–9]. Control signal shielding is complicated by the fact that they must be fed through the apparatus that isolates the quantum data plane from its surroundings via vacuum, cooling, or both; this necessity limits the types of isolation procedures that are possible. Fortunately, both qubit manufacturing faults and signal crosstalk defects are predictable and change slowly with the system's mechanical setup. The effects of these slowly changing errors can be reduced by using control pulse shapes that reduce the qubit's dependence on these factors, as well as through periodic system calibration, if there is a mechanism to measure these errors and software to adjust the control signals to drive these errors to zero (system calibration). Because every control signal has the potential to interact with every other control signal, the number of measurements and computations required to achieve this calibration more than doubles as the system's qubit count doubles [7–11].

8.2.3 Processor Plane and Host Control

Processor plane: This term could refer to a specific layer or level within a processor architecture. Processors typically consist of multiple layers or levels, such as instruction fetch, decode, execution, and memory access. These layers work together to perform various operations on data. The "processor plane" might refer to one of these layers or a specific stage in the processor's pipeline [6–11].

Host control processor: In some systems, especially embedded systems or complex computing systems, there may be a separate processor dedicated to managing and controlling the overall system. This processor is often referred to as the "host control processor." It handles tasks such as system initialization, resource management, communication with external devices, and overall system coordination. It acts as the main controller or supervisor for the entire system [4–9].

8.2.4 Qubit Technologies

In the realm of quantum computing, qubits play a vital role. Qubits, short for quantum bits, are the fundamental units of information in quantum systems. Unlike classical bits that can exist in either a 0 or a 1 state, qubits can exist in a superposition of both states simultaneously, thanks to the principles of quantum mechanics [9–13].

Qubits can be realized using a number of physical systems such as superconducting circuits, trapped ions, topological states, or even photons. These systems provide a platform for manipulating and measuring the quantum states of qubits [8].

By harnessing the properties of qubits, such as superposition and entanglement, quantum computers have the potential to perform certain computations exponentially faster than classical computers. This has significant implications for solving complex problems in areas such as cryptography, optimization, material science, drug discovery, and more [7].

Technologies in qubits are continuously advancing, with ongoing research and development efforts aimed at improving qubit coherence times, reducing errors, increasing the number of qubits in a system, and enhancing overall quantum computational capabilities. Qubit technologies are at the forefront of these advancements, driving innovation and pushing the boundaries of what is possible in quantum computing [6–11].

8.3 TYPES OF QUANTUM COMPUTER

8.3.1 Quantum Annealer

A quantum annealer is a type of quantum computing device designed to solve optimization problems. It is a specialized form of quantum computer that leverages quantum mechanical effects to find the optimal solution among a large number of possibilities. Quantum annealing is based on the concept of adiabatic quantum computation. The annealing process starts with a system of qubits representing the problem to be solved. The qubits are initially placed in a simple and known configuration, and the system is prepared in its ground state. Then, the system is slowly evolved into a final configuration that encodes the solution to the problem [4–10].

During the evolution, the system explores a wide range of configurations and attempts to find the lowest energy state that corresponds to the optimal solution. The goal is to find the configuration that minimizes the objective function associated with the optimization problem [5–7].

Quantum annealers have been particularly useful in solving certain types of optimization problems, such as the traveling salesman problem, graph colouring, and portfolio optimization. They are especially well-suited for solving problems that can be mapped onto the Ising model or the more general quadratic unconstrained binary optimization (QUBO) problems. One of the most well-known quanta annealer implementations is the D-Wave Systems' quantum annealing machine. It utilizes superconducting qubits and operates at very low temperatures to maintain the quantum coherence. However, it's important to note that quantum annealers are not general-purpose quantum computers and have limitations on the types of problems they can effectively solve [22–24].

Quantum annealers are an active area of research and development, and their potential applications continue to be explored in a number of domains, including optimization, machine learning, and scientific simulations.

8.3.2 Analogue Quantum Annealer

An analogue quantum simulator is a device or a system that emulates the behaviour of quantum systems to simulate and study quantum phenomena. It is designed to mimic the dynamics of quantum systems using classical components that can effectively simulate quantum behaviour. Analog quantum simulators are used to investigate quantum phenomena that are difficult to study directly or analytically. They can provide insights into the behaviour of complex quantum systems and help researchers understand quantum effects and develop new algorithms and protocols [7–11, 24–26].

Unlike digital quantum computers, which use discrete qubits and perform computations in a digital manner, analogue quantum simulators rely on continuous variables and analogue components to simulate quantum systems. These simulators employ physical systems, such as trapped ions, superconducting circuits, or cold atoms, which can be manipulated to exhibit the quantum behaviour. The behaviour of the quantum system being simulated is typically encoded in the simulator's parameters, such as energy levels, interaction strengths, and coupling constants. By carefully controlling and manipulating these parameters, researchers can emulate the behaviour of quantum systems and study phenomena such as quantum phase transitions, quantum magnetism, or quantum transport [8–13, 26–28].

Analogue quantum simulators have proven useful in a variety of research areas, including condensed matter physics, quantum chemistry, and quantum many-body systems. They allow scientists to explore the behaviour of quantum systems that are difficult to simulate using classical computers or to gain insights that could inform the design of more powerful quantum algorithms or technologies [14–19, 26–28].

8.3.3 Universal Quantum Computer

A universal quantum computer is a type of quantum computer that is capable of executing general-purpose quantum algorithms and solving a wide range of computational problems. Unlike specialized quantum devices such as quantum annealers or analogue quantum simulators, universal quantum computers have the potential to achieve quantum supremacy, meaning that they can solve problems exponentially faster than classical computers for certain applications [28, 29].

The fundamental building block of a universal quantum computer is the qubit (quantum bit), which is the quantum equivalent of a classical binary bit. Qubits can exist in superposition, representing multiple states simultaneously, and can be entangled with one another, creating intricate quantum correlations. These properties enable quantum computers to perform parallel computations and exploit quantum phenomena such as quantum interference and entanglement. Universal quantum computers operate using quantum gates, which are analogous to classical logic gates but act on qubits in a quantum-mechanical manner. By combining various quantum gates, quantum algorithms can be constructed to solve specific problems efficiently.

Universal quantum computers are still in the early stages of development, and significant engineering and technical challenges need to be overcome to achieve large-scale, fault-tolerant quantum computing. However, they hold the promise of revolutionizing fields such as cryptography, optimization, material science, drug discovery, and machine learning by solving problems that are currently intractable for classical computers [25–29].

8.4 QUANTUM BITS

Quantum bits, often referred to as qubits, are the fundamental building blocks of quantum computing. They are the quantum counterpart to classical bits, which represent binary information as 0 or 1. However, qubits possess unique properties that distinguish them from the classical bits and enable quantum computers to perform computations that are exponentially faster than classical computers for certain applications. Unlike classical bits that can only be in a single state at a time, qubits can exist in a superposition of states. This means that a qubit can simultaneously represent both 0 and 1, or any combination thereof, with varying probabilities. Superposition allows quantum computers to perform computations in parallel, exploring multiple possible states simultaneously [18–22].

Another key property of qubits is entanglement. When qubits are entangled, the state of one qubit becomes correlated with the state of another qubit, regardless of the physical distance between them. This correlation is maintained even if the qubits are physically separated, enabling quantum computers to perform operations on the collective state of entangled qubits. Entanglement allows for powerful computational capabilities, such as quantum teleportation and quantum error correction [23].

Developing and maintaining qubits is a significant challenge in quantum computing. Quantum systems are highly sensitive to environmental disturbances and decoherence, which can cause the loss of quantum information. Quantum error correction techniques are being developed to mitigate the effects of decoherence and enable fault-tolerant quantum computation.

Harnessing the power of qubits and developing robust methods for their manipulation and control are at the forefront of quantum computing research. Quantum algorithms, such as Shor's algorithm for factoring large numbers and Grover's algorithm for searching unsorted databases, leverage the unique properties of qubits to solve specific problems exponentially faster than classical algorithms. While practical, large-scale quantum computers are still in the early stages of development, qubits represent the foundation of these future machines. The race to build more qubits, improve their coherence times, and develop error correction methods continues as researchers work towards realizing the full potential of quantum computing.

8.5 TYPES OF QUBITS

8.5.1 Qubit: Superconductor

The key theory behind superconductivity is that the basic charge carriers are pairs of electrons (known as Cooper pairs) rather than the single electron in a normal conductor. In general, superconducting materials are metals, ceramics, organic materials, or hilly doped semiconductors that conduct electricity without trace [3–7].

Several types of superconducting cuts have been used to implement qubits and quantum logic gates with different properties and potential uses. There are three primary types of superconducting qubits, namely flux, charge and phase qubits. Charge qubit is nothing but an assembly of charges, externally applied magnetic flux controlled the state of flux qubit, whereas external bias current controls the phase of Josephson junction in case of phase qubit. The state of qubit is deemed by the number of Couper pars which have unneedled across the junctions [4–9].

The two primary types of superconducting qubits, the *charge qubit* and the *flux qubit*, are directly related to these two variables: charge qubits are associated with the amplitude, while flux qubits are related to the phase.

Charge qubits are built in a small circuit that includes a superconducting island, a Josephson junction, and a superconducting reservoir. The charge states of the superconducting island serve as the basis states in this scenario.

Flux qubits, also known as persistent current qubits, are micrometre-sized circuits that use an external magnetic flux to produce a persistent current flow. Magnetic flux quanta trapped in a superconducting ring are the basic states in this scenario.

8.5.2 Qubit: Quantum Dot

A quantum dot is a nanoparticle formed of silicon, cadmium selenide, cadmium sulphide, germanium, or indium arsenide. A quantum dot is typically an isolated spherical volume with a diameter of one ten-thousandth of a millimetre that is contained inside a solid. The sphere is "locked inside" by a free electron that is not bound within an atom. The surrounding solid is composed of layers of two semiconducting materials (e.g., silicon and germanium) that have been cooled down to a very low temperature [5–11].

The free electron is held in place by electrical fields, even though it is only one-hundredth of a degree above absolute zero. The electron spin can be switched "up" and "down" electrically in this configuration and can thus be used to store one of the smallest units of information (0/1). In theory, small quantum dots can be used to build computers with hundreds of millions of qubits on a single chip.

8.5.3 Qubit: Trapped Ion

"Ions" are just atoms that have lost or gained one or more electrons, giving them an electrical charge. The ion trap is used to construct quantum registers. The number of qubits in the register equals the number of trapped ions. One electron is lost from the calcium ion. As a result, they are positively charged. When ions absorb and emit single photons, they gain and lose kinetic energy. Long coherence periods allow ion traps to store quantum information and support universal quantum computing. When trapped ions are in a superposition condition, they are extremely stable. They benefit from the ability to become entangled with one another [4–9].

The state of an ion can be determined with great precision (>99.9%) by counting the number of photons captured.

8.5.4 Qubit: Photonic

Photons do not have mass or charge. As a result, they do not interact with one another. Photons are excellent candidates for various quantum information processing and communication activities. They have a lengthy coherence duration, have relatively little contact with the environment, move at the speed of light, and can encode numerous degrees of freedom [4–7].

Their horizontal or vertical polarization determines their photonic quantum state. Photonic qubits work equally well at cryogenic and ambient temperatures, have almost little decoherence, and are easily modified using optical components. They are realized by a universal quantum computing architecture based on phase shifters, beam splitters, and photon counters, known as a linear optical quantum computation.

8.5.5 Qubit: Defect-Based

The nitrogen-vacancy (NV) centre in diamond is a well-known spin defect. Recently, the diamond nitrogen-vacancy (NV-1) centre has emerged as a prominent qubit contender in the solid state because it is an individually addressable quantum system that can be initialized, manipulated, and measured with great fidelity at an ambient temperature [4].

8.5.6 Qubit: Topological

Topology is a part of mathematics that describes structures that undergo physical changes such as being bent, twisted, compacted, or stretched while retaining their original features. When used to quantum computing, topological features provide a level of safety that allows a qubit to preserve information regardless of what is going on around it. The topological qubit accomplishes this additional security through two methods: electron fractionalization and ground state degeneracy. Because topological qubits have higher fidelity, fewer qubits are required for error correction, resulting in a significant reduction in the total number of qubits required by the system [4–9].

Researchers in quantum computing seek to employ a certain type of Majorana fermion, known as a Majorana zero mode, as a qubit. The ultimate quantum computing system may be devices based on topological information protection. Majorana bound states (MBS) in Sm-S nanostructures that produce Andreev bound states (ABSS) at the interface between the typical NW semiconductor (Sm) and the superconductor (S) could be one such system. Applying an axial magnetic field along the S-NW device causes the ABSS to shift to zero energy as the magnetic field increases and forms mid gap states [7].

8.5.7 NMR Qubit

Matter is made up of molecules, which are made up of atoms, and each atom contains a nucleus. Nuclei come in a variety of species known as nuclides, which are distinguished by the number of protons and neutrons they contain. The number of protons in an atom's nucleus defines its chemical makeup [5]. For example, all carbon atoms contain six protons in the nucleus, but all hydrogen atoms have a single proton in the nucleus. Some nuclides are magnetic and have a magnetic moment, which means they interact with magnetic fields that are applied. Nuclear magnetic resonance (NMR) detects the nuclei's bulk magnetism, which is the total of all tiny magnetic moments in the sample.

The experimental observation of the resonant absorption of energy by nuclei from radio-frequency sources is known as NMR. When the nuclear magnetic moment associated with a nuclear spin is put in an external magnetic field, the magnetic potential energies of the different spin states are assigned [6]. A radio frequency signal of the appropriate frequency can trigger a transition between spin states in the presence of a static magnetic field that produces a modest amount of spin polarization. Some of the spins are placed in a higher energy state because of this "spin flip."

When the radio frequency signal is turned off, the relaxation of the spins back to the lower state generates a detectable amount of RF signal at the resonant frequency associated with the spin flip. This is known as NMR. Nuclear spin is impervious to stray magnetic fields. NMR is used to control the nuclear spin state.

This robustness is due to the nuclear spin's magnetic moment having a smaller magnitude than the electronic spins. Nuclear magnetic resonance quantum computing (NMRQC) is one of the several proposed ways for building a quantum computer, in which the spin states of nuclei within molecules are used as qubits. There are two options. The first method is to employ the spin characteristics of atoms in specific molecules in a liquid sample as qubits, which is known as Liquid State NMR (LSNMR) [5–9]. The second method employs solid-state NMR (SSNMR) samples, such as a nitrogen vacancy diamond lattice, rather than liquid samples.

Although NMR qubits offer several advantages, they also face certain challenges that limit their widespread adoption. One primary challenge is the limited qubit interconnectivity, as the qubits typically interact indirectly through their mutual interactions with ancillary spins. This restricts the implementation of large-scale quantum algorithms. Researchers are exploring innovative techniques, such as spin-bath engineering and active feedback control, to overcome these limitations and enhance qubit connectivity.

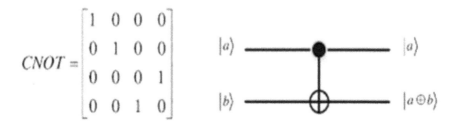

$$CNOT = \begin{bmatrix} 1 & 0 & 0 & 0 \\ 0 & 1 & 0 & 0 \\ 0 & 0 & 0 & 1 \\ 0 & 0 & 1 & 0 \end{bmatrix}$$

Figure 8.1 Matrix of CNOT gate and the corresponding state representation.

Furthermore, efforts are being made to integrate NMR qubits with other qubit platforms to harness the benefits of both technologies. Hybrid systems combining NMR qubits with superconducting qubits or trapped ions show a promise for improving scalability and interconnectivity [8].

Figure 8.1 represents the matrix of CNOT gate and the corresponding state representation.

8.6 APPLICATIONS

In the year 2020, through a research paper, it has been noticed that quantum computing has a vast area of application. As mentioned earlier, quantum computers are quite efficient to solve complex problems that classical computers aren't capable of. Some of the eminent regions of applications of quantum computing are as follows.

8.6.1 Artificial Intelligence and Machine Learning

We know that in the modern-day world, technology has evolved to a huge extent. The main catalyst behind this progress is the introduction of artificial intelligence (AI). The goal of AI is to make an ordinary device smart enough to solve problems as human beings do. The goal of machine learning (ML) is to enable the machine to analyse data and provide an accurate output [11–13]. But with an increase in applications, classical computers can't cope up with the efficiency. This leads to an increase in the usage of quantum computers with better efficiency.

8.6.2 Computation Chemistry

It is a branch of chemistry that deals with computer programs and algorithms to detect the structure and nature of molecules and compounds. This has been regarded as a very useful tool to understand various reaction mechanisms. One of the applications of computational chemistry is green chemistry where the synthesis of a chemical compound is studied based on atom economy, energy efficiency, and so on. Research studies indicate that the inefficiency of classical computers has led to quantum computers with the help of which desired accurate output is obtained. Since the theory states the existence of both 1 and 0 energy states, the mapping of molecular structure is done more precisely [4–9].

8.6.3 Cybersecurity and Cryptography

QKD is a cutting-edge cryptographic technique that utilizes the principles of quantum mechanics to establish secure communication channels. Traditional encryption algorithms rely on the complexity of mathematical problems to protect data, whereas QKD relies on the fundamental principles of physics. By leveraging quantum properties such as superposition and entanglement, QKD provides a means to exchange encryption keys with unconditional security. Quantum computers can enhance QKD protocols, enabling longer key distances, higher key generation rates, and improved overall security. Quantum computing's inherent parallelism and computational power make it well-suited for solving complex optimization problems. In the realm of cybersecurity, quantum algorithms can be employed to enhance tasks such as intrusion detection, malware analysis, and network optimization [12]. By harnessing quantum algorithms, security analysts can efficiently analyse vast amounts of data, identify patterns, and detect anomalies, thereby improving threat detection and response.

8.6.4 Weather Forecasting

In recent studies, it has been noticed that the process of analysing weather conditions by classical computers takes a longer time duration than the weather itself does to change. But a quantum computer's ability to fetch vast amounts of data, in a short period of time, could indeed lead to a well-developed weather system model, allowing scientists to predict the changing weather patterns in no time and with exceptional accuracy—something that is required for the current time when the world is going under a climatic change. Quantum computing can contribute to addressing energy optimization challenges, such as optimizing energy distribution, minimizing energy consumption, and improving renewable energy integration. Furthermore, quantum simulations can aid in climate modelling, facilitating more accurate predictions of weather patterns, climate change impacts, and the development of sustainable energy strategies [14–19].

These are just a few examples of the wide-ranging applications of quantum computing. As the field continues to advance, we can expect further discoveries and innovative solutions across multiple disciplines, transforming industries and shaping the future of technology.

8.7 COMPARISONS OF QUANTUM COMPUTING APPLICATIONS

Quantum computation is essentially a form of analogue computation. Over the past years, there are a number of algorithms of quantum computing that establish its application in different fields such as AI and ML, computational chemistry, cybersecurity and cryptocurrency, drug design, finance modelling, and so on. A comparison of all the algorithms of quantum computing is stated next [4–7].

8.7.1 Margolus and Toffoli Gates

In the study of logic circuits, Toffoli Gate invented by the famous scientist in the year of 1980 is a type of reversible logic gate from which any reversible classical circuits can be derived. It is a universal gate that in other terms is referred to as "controlled-controlled-not" gate whose working principle (truth table) is like the NOT gate.

In a Toffoli gate, 3-bit inputs and outputs are considered, wherein if any two inputs are set to 1, the third bit gets reversed. A study of reversible gates has begun since the 1960s with the research that it does not dissipate much heat [9].

The Toffoli gate is universal; this means that for any Boolean function f(x1, x2, xm), there's a circuit conforming to Toffoli gates that takes x1, x2, xm, and some redundant bits set to 0 or 1 to labours x1, x2, xm, f(x1, x2, xm), and some redundant bits (called scrap). A NOT gate, for illustration, can be constructed from a Toffoli gate by setting the three input bits to {a, 1, 1}, making the third affair bit (1 XOR (a AND 1)) = NOT a; (a AND b) is the third affair bit from {a, b, 0}. Basically, this means that one can use Toffoli gates to make systems that will perform any Boolean function calculation in a reversible manner.

Any reversible gate can be enforced on quantum computers, and hence, the Toffoli gate is also a quantum operator. But the Toffoli gate can't be used for universal quantum calculation, though it does mean that quantum computers can apply all possible classical calculations. The Toffoli gate must be enforced along with some innately quantum gate(s) in order to be universal for calculation. In fact, any single-qubit gate with real portions that can produce a nontrivial amount of state suffices [13]. A Toffoli gate grounded on quantum mechanics was successfully realized in January 2009 at the University of Innsbruck, Austria. While the perpetration of an N-qubit Toffoli with a circuit model requires 2n CNOT gates, the best-known upper set daises at 6n—12 CNOT gates. It has been suggested that trapped ion quantum computers may be suitable to apply an N-qubit Toffoli gate directly. The operation of numerous body commerce could be used for the direct operation of the gate in trapped ions, Rydberg titles, and superconducting circuit executions. Following the dark-state manifold, Khazali-Mølmer Cn-NOT gate operates with only three beats, departing from the circuit model paradigm [12].

8.7.2 Deutsch-Jozsa Algorithm

Proposed by the famous scientists David Deutsch and Richard Jozsa in the year of 1992, it is a type of deterministic quantum algorithm. In this algorithm, certain improvements were implemented by Richard Cleve, Artur Ekert, Chlara Macchaiavello, and Michele Mosca in 1998. In spite of having little practical applications, it is considered one of the primary examples of quantum algorithm, which has an efficiency rate exponentially faster than other classical algorithms [11].

This algorithm was specially designed for a black-box problem that can be solved by a zero-error quantum computer. On the contrary, an exponential number of queries are required by the deterministic classical computer to solve such black-box problems. The Deutsch-Jozsa algorithm is a quantum algorithm that showcases the power of quantum computing in solving a specific type of problem known as the "oracle problem." It was developed by David Deutsch and Richard Jozsa in 1992 and provides a clear example of how quantum computers can outperform classical computers in certain scenarios [21].

The problem addressed by the Deutsch-Jozsa algorithm is known as the "black-box" problem. Classically, given a black box that implements a Boolean function f(x), the goal is to determine whether the function is constant (returns the same value for all inputs) or balanced (returns different values for half of the inputs and the same value for the other half).

The classical approach to solving this problem requires evaluating the function for multiple inputs to determine its behaviour [15]. However, the Deutsch-Jozsa algorithm can solve this problem with just a single query, providing an exponential speedup.

Here's an overview of how the Deutsch-Jozsa algorithm works [18–22]:

    i. Initialization: The algorithm starts with n qubits initialized in the $|0\rangle$ state and one ancillary qubit initialized in the $|1\rangle$ state.

    ii. Superposition: A Hadamard gate is applied to all the n qubits, creating a superposition of all possible input states.

    iii. Query the oracle: The function f(x) is evaluated by applying an oracle gate, which performs a controlled-negation operation based on the hidden nature of the function. This gate flips the phase of the state $|x\rangle$ if the corresponding bit of the hidden string is 1.

    iv. Apply Hadamard gates: Hadamard gates are applied to all the n qubits again.

    v. Measurement: The n qubits are measured, resulting in the determination of whether the function f(x) is constant or balanced.

By utilizing quantum parallelism and interference effects, the Deutsch-Jozsa algorithm can determine the nature of the function f(x) with just a single query, providing an exponential speedup compared to classical methods that require multiple queries.

The significance of the Deutsch-Jozsa algorithm lies in its demonstration of the power of quantum computing in solving the oracle problem more efficiently than classical computers. While the problem itself may seem simple, the algorithm showcases the underlying principles of quantum algorithms and their potential impact on various fields, including cryptography and optimization.

Overall, the Deutsch-Jozsa algorithm serves as a milestone in quantum computing, highlighting the advantage of leveraging quantum principles to solve computational problems exponentially faster than classical approaches.

8.7.3 Bernstein-Vazirani Algorithm

The Bernstein-Vazirani algorithm is a quantum algorithm that demonstrates the power of quantum computers in solving certain types of problems exponentially faster than classical computers. It was developed by Ethan Bernstein and Umesh Vazirani in 1992 and is a prime example of how quantum algorithms can provide a speedup over classical algorithms [12–14].

The problem addressed by the Bernstein-Vazirani algorithm is known as the "hidden string" problem. *In classical terms, it involves guessing a binary string of 0s and 1s based on a series of queries. Specifically, given a function f(x) that takes an n-bit binary input x and returns the bitwise dot product of x and a hidden n-bit binary string s, the goal is to determine the hidden string s.*

Classically, this problem requires n queries to the function f(x) to identify all the bits of the hidden string. However, the Bernstein-Vazirani algorithm can solve this problem with just a single query, providing an exponential speedup.

Here's an overview of how the Bernstein-Vazirani algorithm works [13–17]:

a. Initialization: The algorithm starts with n qubits initialized in the $|0\rangle$ state and one ancillary qubit initialized in the $|1\rangle$ state.
b. Superposition: A Hadamard gate is applied to all the n qubits, creating a superposition of all possible input states.
c. Query the function: The function f(x) is evaluated by applying an oracle gate, which performs a controlled-phase operation based on the hidden string s. This gate flips the phase of the state $|x\rangle$ if the corresponding bit of s is 1.
d. Measure the result: The n qubits are measured, resulting in the binary representation of the hidden string s.

By utilizing quantum parallelism and interference effects, the Bernstein-Vazirani algorithm efficiently extracts the hidden string s with a single query to the function f(x). This is in stark contrast to the classical approach that requires multiple queries to identify the hidden string bit by bit.

The significance of the Bernstein-Vazirani algorithm lies in its demonstration of quantum computing's ability to achieve exponential speedups over classical algorithms for certain problems. While the hidden string problem may appear simple, the algorithm showcases the underlying power of

Table 8.1 Tabular Comparison of the Success Probabilities of Different Algorithms.

Connectivity	Star-shaped			Fully connected		
Hardware	**Superconducting**			**Ion Trap**		
Success %	**Obs**	**Rand**	**Sys**	**Obs**	**Rand**	**Sys**
Margolus	74.1(7)	82	75	90.1 (2)	91	81
Toffoli	52.6(8)	78	59	85.0 (2)	89	78
Bernstein-Vazirani	72.8(5)	80	74	85.1(1)	90	77

quantum algorithms and their potential impact on cryptography, optimization, and other fields [23–27].

Overall, the Bernstein-Vazirani algorithm serves as a cornerstone in quantum computing, highlighting the advantage of harnessing quantum principles to tackle computational problems more efficiently than classical approaches.

8.8 RECENT WORKS

The past few years have seen a dramatic increase in the interest in quantum computing, both from a theoretical and from a practical perspective. This is motivated by the fact that quantum computers can potentially solve certain problems much faster than classical computers. However, there are still many challenges that need to be addressed before quantum computers can be widely used. In this chapter, we focus on some recent trends and challenges in quantum computing based on AI [23–27].

8.8.1 Case Study 1

According to a study, organized by the Massachusetts Institute of Technology, a computing architecture was proposed by the researchers that had the ability to permit high-fidelity communication among the superconductors in a quantum processor. In this technique, it was observed that a photon (quanta) can transmit data in a direction predetermined by the user. In a paper named "A study on the basics of Quantum Computing" by Montreal University, it was recorded that the foundation of quantum computing is well demonstrated while quantum algorithms along with logic gate operations, error correction, and control of decoherence play an important role in the growth of quantum computing. *Quantum decoherence* is the loss of quantum coherence [22]. In quantum mechanics, particles such as electrons are described by a wave function, a mathematical representation of the quantum state of a system; a probabilistic interpretation of the wave function is used to explain different quantum effects [4]. If there exists a definite phase relation between different states, the system is said to be coherent. Decoherence was first introduced by a German physicist named H. Dieter Zeh in the year of 1970. It is a loss of information by a system into the environment, assuming that every system is loosely coupled with the energy state [2–5].

8.8.2 Case Study 2

On the contrary, we get to know that in logic gate operations, quantum computers operate using qubits, not bits. Unlike traditional bits, which can only be 0 or 1, a qubit can exist in a "superposition" of 0 and 1. This ability to exist in multiple states at once gives quantum computers tremendous power. But a qubit is useless unless you can use it to carry out a quantum calculation. And these quantum calculations are achieved by performing a series of fundamental operations, known as quantum logic gates. There are a lot of types of quantum gates. There are single-qubit gates, which can flip a qubit from 0 to 1, as well as allow superposition states to be created [5–8].

As quantum computing continues to develop, so too do the potential applications for quantum computers. One area of recent research is using quantum computers for AI. There are many potential advantages to using quantum computers for AI, including the ability to process massive amounts of data and the ability to find solutions to complex problems. However, there are also challenges that need to be addressed before quantum computers can be used for AI on a large scale. In this chapter, we explore some of the recent trends and challenges in quantum computing based on AI [4–9].

8.8.3 Case Study 3

One trend in quantum computing is the development of hybrid systems that combine classical and quantum computing. These hybrid systems can take advantage of the strengths of both types of computing. For example, a classical computer can be used to pre-process data before it is fed into a quantum computer. This can help reduce the size of the data set that needs to be processed by the quantum computer, which can save time and resources. Hybrid systems can also be used to post-process data that has been generated by a quantum computer. This can help improve the accuracy of results and make them more useful for practical applications [5–6].

Another trend is the use of ML algorithms on quantum computers. ML is a type of AI that allows computers to learn from data without being explicitly programmed. Quantum computers have the potential to vastly outperform classical computers at ML tasks, such as pattern recognition and classification. This could lead to breakthroughs in fields such as healthcare and finance.

- The old quantum theory's emergence of an array of seemingly unrelated problems in diverse areas such as statistical physics, radiation theory, and spectroscopy; how it was applied to an increasing number of problems; and how the physics community came to recognize its limitations.

- The genesis of modern quantum mechanics in the period around 1925, its conceptual development, the interplay with experiment, its socio-cultural and institutional context, as well as the debates about the different mathematical formulations of the theory (matrix and wave mechanics, transformation theory) and their physical interpretation (statistical interpretation, uncertainty principle).
- The acceptance of quantum mechanics as a new basis for physics (atomic, molecular, nuclear, and solid-state) and parts of chemistry; the elaboration of its mathematical formalism; the establishment of the dominant Copenhagen interpretation and the emergence of critical responses; and subsequent developments up to the present, including the ability to produce and control phenomena that until recently existed only as a theoretical speculation [11–14, 19].

8.8.4 Some of the Recent Works on Quantum Computing

1. Quantum supremacy: In 2019, Google's research team claimed to have achieved "quantum supremacy," demonstrating that a quantum computer could perform a specific task faster than the most powerful classical supercomputers. They used a 53-qubit quantum processor to solve a random sampling problem, showcasing the potential of quantum computing in tackling complex computational challenges [6].

2. Error correction: One of the critical challenges in quantum computing is dealing with errors caused by environmental noise and imperfections in quantum hardware. Recent research has focused on developing robust error correction techniques to enhance the reliability and stability of quantum computations. Advances in error correction codes and error mitigation strategies are crucial for realizing fault-tolerant quantum computers [7].

3. Quantum ML: The intersection of quantum computing and ML has gained significant attention. Researchers are exploring ways to leverage quantum algorithms and quantum data structures to enhance a number of ML tasks, including classification, regression, and clustering. Quantum ML algorithms aim to exploit the unique properties of quantum computing to improve computational efficiency and enable new insights in data analysis [8].

4. Quantum algorithms for optimization: Optimization problems, such as portfolio optimization, scheduling, and resource allocation, are of immense practical importance across the industries. Recent works have focused on developing quantum algorithms that can outperform classical algorithms in solving optimization problems. Quantum

optimization algorithms, such as the Quantum Approximate Optimization Algorithm (QAOA) and the Quantum Variational Eigen solver (QVE), aim to find optimal solutions more efficiently [9].

5. Quantum cryptography and security: With the advent of powerful quantum computers, there is a growing need for quantum-resistant cryptographic techniques. Researchers have made progress in developing post-quantum cryptographic algorithms that can withstand attacks from quantum computers. Quantum key distribution protocols and quantum random number generation techniques are also being explored to enhance the security of communications and data [10].

8.9 FUTURE WORKS AND CONCLUSION

In the distant future, quantum computing will enable industries to overcome problems that were once impossible to solve. For example, pharmaceutical companies could benefit from accelerating the discovery of newer drugs, and transportation companies can have their logistics. Problems optimized, finance companies will be able to create newer trading strategies, oil companies can calculate how atoms and molecules could be configured to help protect equipment from corrosion, and airlines can seek an optimal pathway in storing spare parts at airports.

Quantum computers can simulate the behaviour of matter down to the molecular level, as stated by an article published by the Massachusetts Institute of Technology (MIT).

The batteries of electric vehicles can be improved with the help of quantum computers. For instance, automobile manufacturers such as Volkswagen and Daimler are using quantum computers to simulate the chemical composition of electric vehicle batteries.

Quantum computing is also used by pharmaceutical companies to analyse and compare compounds that could lead to the creation of new drugs.

Since quantum computers can navigate vast numbers of potential solutions extremely fast, they can be used for optimization problems. For instance, Airbus, a European multinational aerospace corporation, is using quantum computing to help calculate the most fuel-efficient ascent and descent paths for aircraft.

The MIT article states that Volkswagen has unveiled a service that minimizes congestion by calculating the optimal routes for buses and taxis in cities.

D-Wave Systems, a Canadian quantum computing company, has built a quantum computer, which the firm says is the first and only quantum computer designed for business use.

According to a statement released by the company, the quantum computer, known as Advantage quantum system, has been designed with a new processor architecture with over 5,000 qubits and 15-way qubit connectivity. The firm claims that this enables companies to solve their largest and most complex business problems.

REFERENCES

[1] Woolnough, A.P., Hollenberg, L.C., Cassey, P., Prowse, T.A., 2023. Quantum computing: A new paradigm for ecology. Trends in Ecology & Evolution 38(8).

[2] Mitchell, C.J., 2020. The impact of quantum computing on real-world security: A 5G case study. Computers & Security 93, 101825.

[3] Hu, F., Lamata, L., Sanz, M., Chen, X., Chen, X., Wang, C., Solano, E., 2020. Quantum computing cryptography: Finding cryptographic Boolean functions with quantum annealing by a 2000 qubit D-wave quantum computer. Physics Letters A 384(10), 126214.

[4] McGeoch, C.C., 22 January 2020. Theory versus practice in annealing-based quantum computing. Theoretical Computer Science 816.

[5] Ajagekar, A., You, F., 5 December 2020. Quantum computing assisted deep learning for fault detection and diagnosis in industrial process systems. Computers & Chemical Engineering 143, 107119.

[6] Arute, F., et al., 2019. Quantum supremacy using a programmable superconducting processor. Nature 574, 505–510.

[7] Bergholm, V., Izaac, J., Schuld, M., Gogolin, C., Killoran, N., 2018. *Pennylane: Automatic differentiation of hybrid quantum-classical computations.* arXiv preprint. https://arxiv.org/abs/1811.04968.

[8] Biamonte, J., Wittek, P., Pancotti, N., Rebentrost, P., Wiebe, N., Lloyd, S., 2017. Quantum machine learning. Nature 549, 195.

[9] Chollet, F., et al., 2018. *Keras: The python deep learning library.* Astrophysics Source Code Library. http://ui.adsabs.harvard.edu/abs/2018ascl.soft06022C/abstract.

[10] Feynman, R.P., 1982. Simulating physics with computers. International Journal of Theoretical Physics 21, 467–488.

[11] Fingerhuth, M., Babej, T., Wittek, P., 2018. Open source software in quantum computing. PLOS One 13, e0208561.

[12] Gamel, O., 2016. Entangled Bloch spheres: Bloch matrix and two-qubit state space. Physical Review A 93, 062320.

[13] Grant, E., Benedetti, M., Cao, S., Hallam, A., Lockhart, J., Stojevic, V., Green, A.G., Severini, S., 2018. Hierarchical quantum classifiers. NPJ Quantum Information 4, 65.

[14] Grover, L.K., 1996. *A fast quantum mechanical algorithm for database search.* Proceedings of the Twenty-Eighth Annual ACM Symposium on Theory of Computing, ACM, New York, NY, pp. 212–219.

[15] Havlíček, V., Córcoles, A.D., Temme, K., Harrow, A.W., Kandala, A., Chow, J.M., Gambetta, J.M., 2019. Supervised learning with quantum-enhanced feature spaces. Nature 567, 209.

[16] Kingma, D.P., Ba, J., 2014. *Adam: A method for stochastic optimization.* arXiv preprint. https://arxiv.org/abs/1412.6980.

[17] LeCun, Y., Bengio, Y., Hinton, G., 2015. Deep learning. Nature 521, 436–444.

[18] Nielsen, M.A., Chuang, I.L., 2011. *Quantum computation and quantum information*, 10th anniversary edition. Cambridge University Press, New York, NY.

[19] Preskill, J., 2018. Quantum computing in the NISQ era and beyond. Quantum 2, 79.

[20] Reinsel, D., Gantz, J., Rydning, J., 2018. *The digitization of the world: From edge to core*. International Data Corporation, Framingham.

[21] Rumelhart, D.E., Hinton, G.E., Williams, R.J., 1986. Learning representations by back-propagating errors. Nature 323, 533–536.

[22] Schuld, M., Bergholm, V., Gogolin, C., Izaac, J., Killoran, N., 2019. Evaluating analytic gradients on quantum hardware. Physical Review A 99, 032331.

[23] Schuld, M., Fingerhuth, M., Petruccione, F., 2017. Implementing a distance-based classifier with a quantum interference circuit. EPL: Europhysics Letters 119, 60002.

[24] Schuld, M., Petruccione, F., 2018. *Supervised learning with quantum computers*, vol. 17. Springer International Publishing, Cham.

[25] Steane, A., 1998. Quantum computing. Reports on Progress in Physics 61, 117–173.

[26] Sutskever, I., Martens, J., Dahl, G., Hinton, G., 2013. On the importance of initialization and momentum in deep learning. International Conference on Machine Learning 1139–1147.

[27] Tacchino, F., Macchiavello, C., Gerace, D., Bajoni, D., 2019. An artificial neuron implemented on an actual quantum processor. NPJ Quantum Information 5, 26.

[28] Wolberg, W.H., Street, W.N., Mangasarian, O.L., 1995. *UCI machine learning repository*.

[29] Zhou, L., Pan, S., Wang, J., Vasilakos, A.V., 2017. Machine learning on big data. Neurocomputing 237, 350–3611.

Chapter 9

Quantum Microwave Engineering

A New Application Area of Quantum Computing

Pampa Debnath, Arpan Deyasi, and Siddhartha Bhattacharyya

9.1 INTRODUCTION

Modern microwave engineering and quantum computing have a similar progenitor in the early research that produced radar and related technologies [1]. Since quantum mechanics describes that the many of the basic processes involve the interaction of microwaves with atoms or molecules as well as their individual charge and spin states, many of the fundamental mechanisms underlying the creation, transmission, absorption, and recognition of microwave energy [2] were known to be directed by this theory at the time. An excellent illustration of the long-distance interaction of microwave engineering and quantum method is the employment of the radar technology and techniques developed during World War II to discover solid-state nuclear magnetic resonance.

Resonant interactions between microwave photons and other quantum objects [3], such as the quantum two-level systems (known as qubits), serve as the fundamental units of quantum computing, according to the quantization of microwave power. The exact management and delivery of quantum states are now made possible by the widespread utilization of microwave expertise across different quantum platforms.

Quantum computing algorithms can be applied to simulate the electromagnetic structures [1], precisely the electromagnetic field states [4], when they are equivalently represented by qubits. Time parallelism is assumed for the simulation of a large number of identical structures, in order to facilitate the linear superposition principle [5]. However, the demand for efficient algorithms rises as the number of structures to be simulated in parallel becomes extremely large, and therefore, the superposition of structures should be invoked in that algorithm [4].

Earlier research started with the Transmission Line Matrix (TLM) method using Hilbert Space Transformation [6]; however, only perfect electrical conductors and free space are considered for using this method. The need of time evaluation operator becomes the significant area of research, and the first paper reported it using CNOT gates [7]. Later, it was modified with the inclusion of single-qubit and Toffoli gates [4].

DOI: 10.1201/9781003373117-9

Algorithms so far applied in the domain of quantum microwave are in a nascent state, and a few researchers highlighted the optimal quantum microwave control [8]. Quantum parallelism is one of the biggest challenges for proceeding simultaneous solutions of electromagnetic structures, and thus, optical control is proposed [8]. Neural network–driven reinforcement learning method [9–10] is one of the prime techniques for this purpose, owing to its variety of applications in the domain of quantum architecture [11], quantum thermodynamics [12], quantum circuit [13], adaptive quantum metrology [14], and so on. However, for physical problems with the closed loop, the algorithm essentially fails as it exhibits very poor performance. Alternative solutions are searched in this context, and hyperparameter tuning tests become the possible solution.

9.2 QUANTUM MICROWAVE PROPAGATION

Like the propagation of conventional electromagnetic waves, microwave propagation in quantum regime can be sub-categorized as the guided wave transmission and the unguided wave transmission. In the following sub-sections, both the categories are highlighted briefly.

9.2.1 Guided Propagation

Coaxial cables made of guided superconducting niobium-titanium are often utilized nowadays for low-loss microwave signals. These cables generally display very low absorption losses and have a characteristic impedance of 50. The surface quality of the superconducting material mostly controls these losses and the loss tangent of the appropriate dielectrics, such as polytetrafluoroethylene (PTFE). These cables provide a versatile, reliable, and an easily accessible means to connect numerous equipment in a very low temperature. Using different rigid waveguides built of aluminium or niobium is an alternate method for low-loss guiding of quantum microwave signals [15]. In these systems, electromagnetic fields are greatly diluted due to increased waveguide diameters and resulting inner volumes, which minimizes coupling with different dissipative channels [16].

9.2.2 Non-Guided Propagation

The normal situation for microwave communications in free space is non-guided propagation. This occurs when a quantum state is broadcast from the cryostat using a "quantum antenna" [17–18], a device that can broadcast quantum microwave states to the free space with adjustable directivity [15]. This technology accomplishes impedance matching between the cryostat and the free space by creating a finite cavity. These two mediums,

that is, a cryostat with 50 Ω impedance and a free space with 377 Ω impedance, have extremely different impedances, which affect the signal propagation. An appropriate antenna is, therefore, required to reduce the unwanted
reflection of signals, which may be caused due to the impedance mismatch of
two different mediums. This impedance mismatch problem for microwave
two-mode squeezed (TMS) states was investigated, and it was discovered
that entanglement conservation was closely connected to the reflectivity
and, consequently, to the structure of the antenna [19]. As a result, limiting
reflections and eliminating the effects of thermal radiation should also be
priorities for the quantum antennas.

9.3 QUANTUM COMPUTING WITH QUBITS

Quantum bits, or qubits, are the basic message transporters in a quantum
computer, such as the bits with logic used in a classical (non-quantum) computer. To illustrate this, the section is sub-divided into basics of qubit, followed by the qubit working as a resonator.

9.3.1 Qubit Basics

According to Figure 9.1a, a qubit is nothing but a quantum mechanical
system that has eigenstates with two energy levels, which we denote by the
letters |0| and |1|, and associated energy eigenvalues E_0 and E_1, which take
the value E. A qubit is a particle with spin-1/2, in a magnetic field, like an
electron, but there are a lot of alternative ways a qubit could be realized.

Figure 9.1b shows the transition between two quantum states by qubit
and the corresponding energy difference. The generated oscillation between
them is exhibited in Figure 9.1c. The corresponding probability distribution
due to Rabi oscillation is shown under the resonance condition.

Figure 9.2 depicts the oscillations displayed by the qubit when it is excited
by the source. The upward/downward transition frequency is ω_{01} (as shown
in Figure 9.1). If the thermal energy is less than the transition energy, then

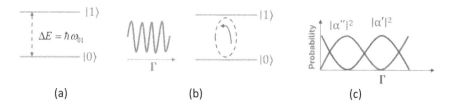

(a)   (b)   (c)

Figure 9.1 Basics of qubits: (a) two quantum states and |0) and |1) making qubit
with ΔE energy difference; (b) oscillation between two quantum states
for resonance excitation; (c) probability distributions for Rabi oscillation under continuous wave resonant driving [2].

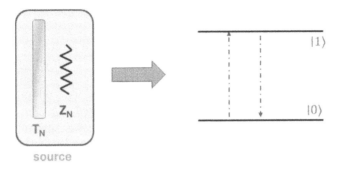

Figure 9.2 Oscillations between two quantum states exhibited by the qubit when excited by the source (characterized by impedance Z_N and temperature T_N) [2].

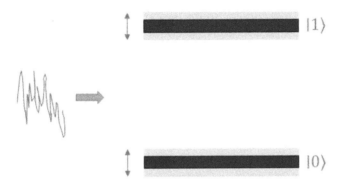

Figure 9.3 Effect of an environmental/external excitation on shifting of qubit frequency [2].

downward transition will dominate, whereas for the reverse case, upward will be the domination one.

Before considering the specific physical recognition of qubits and their ties to microwave technology, we first outline the fundamental traits and activities of qubits. The two states, '0' and '1' are utilized in traditional computing, and they are equivalent to the $|0|$ and $|1|$ states. These are considered as computational base states.

Figure 9.3 exhibits the shift of qubit frequency due to an external excitation, which may be generated by the applied signal or by pure environmental excitation. The change may be small in magnitude; however, its rapid fluctuation causes change, which causes spreading of both the quantum states, and naturally, transition frequency is changed. This inherently causes change in transition energy.

The immediate status of a qubit ψ can be expressed as a linear amalgamation of the eigenstates with two energy states and having intricate amplitudes

0 and 1, just like the electromagnetic (EM) field can be divided into a direct amalgamation of orthogonal modes α' and α''.

In contrast to how traditional bits behave, which is limited to one status at a time, the qubit may be in both positions |0| and |1| simultaneously due to the quantum-mechanical structure of the state, an event recognized as the superposition. However, the qubit status is supposed to "collapse" to only one of its eigenstates when the qubit status is measured.

More than one qubit is required for successful computing on quantum computers. There are 2N base positions of the structure for N-qubits, ranging from |00 . . . 00| to |11 . . . 11|. N-qubits with the quantum state ψ can be a linear amalgamation of any or even all of the 2N base status at the equivalent time, with analogous composite probability amplitudes 00 . . . 00, 00 . . . 01, and 11 . . . 11. N classical bits, in contrast, can only ever be in one of the 2N base states at any given moment [15]. These amplitudes satisfy the normalizing requirement 2N–1 k = 0 |k | 2 = 1, where k indexes the 2N various bit strings corresponding to the base states. This is so because, in terms of the likelihood that a measurement would be successful, these amplitudes have a physical significance.

There is a potential for a considerably greater computing capacity since N-qubits with 2N-dimensional state space can store exponentially more data than that of N traditional bits. However, the measured fall down phenomena means that, at the wrapping up of an algorithm, only N bits of message, chosen probabilistically by the measurement procedure, may be recovered from the quantum status. When every quantum basis status is assigned to a port, the behaviour of an N-qubit algorithm of quantum computing is equivalent to the scattering matrix for a 2N-port microwave device, which is passively lossless. Quantum algorithms regulate the status of each qubits and produce entanglement between qubits as they run, resulting in the required interference between probability amplitudes. In experiments, these activities typically include the use of microwave signals or heavily rely on microwave technology. Microwave signals are also used for measurements in several qubits, referred to as the reliability, which can be used to describe how well the qubit state preparing, organizing, and measuring processes operate. A reliability of |1| indicates a flawless implementation, whereas |0| denotes total failure. The reliability may be regarded of as reflecting how near the laboratory procedure for a task is as per its perfect theoretical description, defined by quantum entanglement. Implementation mistakes might result from the existence of dissipation of energy, unwanted signal or noise, drift of qubits between adjacent quantum states, or wrong calibration, even at the sub-nano level, which ultimately lowers the reliability to less than the unit value. In order to compute the deficit, the error rate is estimated, which is nothing but a measure of how much below one the reliability of a task is. Research is now being done on methods for quickly and precisely evaluating reliabilities and rates of error, especially in bigger quantum computers.

9.3.2 Qubits Operated as Resonators

Qubits are incredibly harmonic (or nonlinear), in comparison to conventional linear resonators. The transition of qubit can be made possible from the lowest quantum level, defined as "ground" $|0\rangle$ state to the "excited" $|1\rangle$ state, using a resonant continuous wave drive tone; however, an added excitation is not at all feasible since resonant upper energy states do not exist practically, and therefore, further transitions are never possible [20]. Thus, applying the driving tone repeatedly can only result in the qubit returning to $|0\rangle$. Following the superposition principle, any qubit state may be assumed as the linear superposition of the ground state $|0\rangle$ and the excited state $|1\rangle$ [21], which generates sustained resonant driving of that qubit under consideration as a result and, consequently, leads to sinusoidal oscillations of the probabilities $|\alpha|^2$ and $|\alpha'|^2$ in time as observed in Figure 9.1c. Such type of oscillations are nomenclature as Rabi oscillations, whose angular frequency is governed by the magnitude of the resonant drive.

The total internal quality factor, symbolized as Q_i of a qubit, as in classical resonators, reflects dissipation brought on by intrinsic loss processes. The quality factors of various ports are amalgamated in Q_i such as for driving ports (Q_d), for coupling ports (Q_c), and for measuring ports (Q_m), which may be used to identify each loss channel. The overall quality factor is computed by using the following formula:

$$Q_{overall}^{-1} = Q_d^{-1} + Q_c^{-1} + Q_m^{-1} + Q_i^{-1}$$

The connection to many sources of dissipation, each of which has an effective noise temperature that is generally (but not always) close to the physical temperature, is one method to conceptualize these loss channels. This is indicated in Figure 9.4.

There is another kind of decoherence that we must consider along with the existing loss owing to the damping effect. When a qubit resonator is excited, it eventually leads to phase decoherence with time. This can be measured when compared with phase output of steady reference oscillator

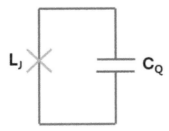

Figure 9.4 Schematic representation of transmon qubit, where L_J represents the Josephson junction and C_Q is the lossless (ideal) capacitor [2].

or standardized with an ideal lossless resonator, because arbitrary fluctuation of resonance frequency leads to that decoherence.

9.4 PHYSICAL REALIZATION OF QUBIT

Implementation of qubits can be realized physically in different methods, much like classical bits—for example, the voltage on a transistor's gate, the spin alignment of a tiny magnetic area on a hard drive, or the reflectivity of a small area of an optical storage media. In this chapter, we concentrate on three prominent physical qubit implementations, namely, semiconductor spins, trapping of ions, and superconducting circuits.

9.4.1 Qubit Trapped Ion

This example may be treated as a sophisticated, highly advanced, and high-reliability quantum technique for the physical realization of the position of qubits in quantum states that are composed of atomic ions confined in extremely high vacuum chambers [22]. These confined ions are designated as linear RF traps [23] applicable to quantum computing. Trapping of these ions becomes possible owing to the combined effect of oscillating the RF electric field and the static electric field under the fundamental condition that both will be in the phase sequence over a very long duration of time, which may range up to several months, depending on specification. The generation of these fields is possible by applying suitable RF potential (for the oscillating filed) and dc potential (for the static field) to the groups of trapping electrodes.

9.4.2 Spin Qubits for Semiconductors

Another possible base for restructurable quantum computing architectures is the freedom from spin degree in materials. Spin qubits are immersed in materials and accompanied by similar atoms, which is quite contrary to the previous paragraph, where ion is trapped in vacuum in single. Many of these atoms can connect with spin qubits (immersed in material) in an unpredictable manner [24]. Luckily, in materials with few nuclear spins like carbon, silicon-germanium, and silicon compounds like diamond, the aforementioned interactions are rather weak.

In general, strongly isolated qubits that exhibit weak interactions with their surroundings are likewise weakly linked to any form of control. Slow timings of quantum gate are a result of poor coupling to regulate fields, which may negate any benefit from the lengthy coherence periods [25]. A balancing act is carried out between the ability to be controlled, device complication, and susceptibility to charge or generating noise. This ultimately produces different alternative states of spin qubits.

9.4.3 Superconducting Qubits

Macroscopic devices have been considered as superconducting qubits that are specified at the implementation level. Lossless capacitors are the major unit block for developing the circuits, along with the Josephson junctions and inductors. A very basic level of representation is shown in Figure 9.4. These circuits exhibit quantum mechanical behaviour in exact coherent states (both time and phase), which is one of the fundamental requirements for the possible manifestation of the quantum processor [26]. Operating temperatures of these circuits are extremely low, approximately of the order of mK. The characteristics of super conductive qubits may be developed similarly to those of conventional circuit architectures since they are built utilizing distributed or/and lumped circuit components. As a result, this technological platform is capable of realizing a diverse set of quantum processors.

9.5 CONCLUSION

In this chapter, a brief theoretical review is carried out on quantum microwave propagation, both in unguided and guided mediums, and the corresponding working of resonator is explained in the light of qubit. The basic features of qubit and its role in devising spin state and superconducting state are briefly summarized. The fundamental features and mode of application areas of qubit will help to make its manifestation for future possible electromagnetic transport in quantum regime.

REFERENCES

[1] J. C. Bardin, D. Sank, O. Naaman, E. Jeffrey, "Quantum Computing: An Introduction for Microwave Engineers", IEEE Microwave Magazine, vol. 21, no. 8, pp. 24–44, 2020

[2] J. C. Bardin, D. H. Slichter, D. J. Reilly, "Microwaves in Quantum Computing", IEEE Journal of Microwaves, vol. 1(1), pp. 403–427, 2021

[3] N. Lauk, N. Sinclair, S. Barzanjeh, J. P. Covey, M. Spiropulu, C. Simon, "Perspectives on Quantum Transduction", Quantum Science and Technology, vol. 5, p. 020501, 2020

[4] S. Sinha, P. Russer, "Quantum Computing Algorithm for Electromagnetic Field Simulation", Quantum Information Process, vol. 9, pp. 385–404, 2010

[5] D. Deutsch, "Quantum Theory, the Church-Turing Principle and the Universal Quantum Number", Proceedings of Royal Society A, vol. 400, pp. 97–117, 1985

[6] P. Russer, M. Krumpholz, "The Hilbert Space Formulation of the TLM Method", International Journal of Numerical Modelling, vol. 6, pp. 29–45, 1993

[7] M. A. Nielsen, I. L. Chuang, Quantum Computation and Quantum Information, 1st ed., Cambridge University Press, Cambridge, 2000

[8] J. Brown, M. Paternostro, A. Ferraro, "Optimal Quantum Control via Genetic Algorithms for Quantum State Engineering in Driven-Resonator Mediated Networks", Quantum Science and Technology, vol. 8, p. 025004, 2023

[9] N. M. P. Neumann, P. B. U. L. deHeer, F. Phillipson, "Quantum Reinforcement Learning: Comparing Quantum Annealing and Gate-Based Quantum Computing with Classical Deep Reinforcement Learning", Quantum Information Processing, vol. 22, p. 125, 2023

[10] D. Dong, C. Chen, H. Li, T. J. Tarn, "Quantum Reinforcement Learning", IEEE Transactions on Systems, Man, and Cybernetics, Part B (Cybernetics), vol. 38(5), pp. 1207–1220, 2008

[11] X. Zhu, X. Hou, "Quantum Architecture Search via Truly Proximal Policy Optimization", Scientific Reports, vol. 13, Art id: 5157, 2023

[12] P. Sgroi, G. M. Palma, M. Paternostro, "Reinforcement Learning Approach to Nonequilibrium Quantum Thermodynamics", Physical Review Letters, vol. 126, no. 2, p. 020601, 2021

[13] H. Fan, C. Guo, W. Luk, "Optimizing Quantum Circuit Placement via Machine Learning", DAC 22: Proceedings of the 59thACM/IEEE Design Automation Conference, pp. 19–24, 2022

[14] P. Palittapongarnpim, P. Wittek, E. Zahedinejad, S. Vedaie, B. C. Sanders, "Learning in Quantum Control: High-Dimensional Global Optimization for Noisy Quantum Dynamics", Neurocomputing, vol. 268, pp. 116–126, 2017

[15] M. Casariego, E. Z. Cruzeiro, S. Gherardini, T. Gonzalez-Raya, R. André, G. Frazao, G. Catto, D. Datta, K. Viisanen, J. Govenius, M. Prunnila, M. Möttönen, K. Tuominen, M. Reichert, M. Renger, K. G. Fedorov, F. Deppe, H. van der Vliet, A. J. Matthews, Y. Omar, Y. Fernández, "Propagating Quantum Microwaves: Towards Applications in Communication and Sensing", Quantum Science and Technology, vol. 8, p. 023001, 2023

[16] P. Kurpiers, M. Pechal, B. Royer, P. Magnard, T. Walter, J. Heinsoo, Y. Salathé, A. Akin, S. Storz, J.-C. Besse, S. Gasparinetti, A. Blais, A. Wallraff, "Quantum Communication with Time-Bin Encoded Microwave Photons", Physical Review Applied, vol. 12, p. 044067, 2019

[17] S. Mikki, "Quantum Antenna Theory for Secure Wireless Communications", IEEE 14th European Conference on Antennas and Propagation (EuCAP), 5–20 March 2020, Copenhagen, Denmark

[18] S. Mikki, "Fundamental Spacetime Representations of Quantum Antenna Systems", Foundations, vol. 2, no. 1, pp. 251–289, 2022

[19] G. Y. Slepyan, S. Vlasenko, V. D. Mogilevtsev, "Quantum Antennas", Advanced Quantum Technology, vol. 3, p. 1900120, 2020

[20] A.-B. A. Mohamed, H. Rmili, M. Omri, A-H. Abdel-Aty, "Two-Qubit Quantum Nonlocality Dynamics Induced by Interacting of Two Coupled Superconducting Flux Qubits with a Resonator Under Intrinsic Decoherence", Alexandria Engineering Journal, vol. 77, pp. 239–246, 2023

[21] J. W. Silverstone, R. Santagati, B. Ingelheim, D. Bonneau, M. J. Strain, "Qubit Entanglement Between Ring-Resonator Photon-Pair Sources on a Silicon Chip", Nature Communications, vol. 6, no. 1, pp. 7948, 2015

[22] C. D. Bruzewicz, J. Chiaverini, R. McConnell, J. M. Sage, "Trapped-Ion Quantum Computing: Progress and Challenges Featured", Applied Physics Reviews, vol. 6, p. 021314, 2019

[23] B. Lekitsch, S. Weidt, A. G. Fowler, K. Mølmer, S. J. Devitt, C. Wunderlich, W. K. Hensinger, "Blueprint for a Microwave Trapped Ion Quantum Computer", Science Advances, vol. 3, no. 2, p. e1601540, 2017

[24] G. Burkard, T. D. Ladd, A. Pan, J. M. Nichol, J. R. Petta, "Semiconductor Spin Qubits", Review of Modern Physics, vol. 95, p. 025003, 2023

[25] H. Bluhm, L. R. Schreiber, "Semiconductor Spin Qubits—A Scalable Platform for Quantum Computing?", IEEE International Symposium on Circuits and Systems (ISCAS), 26–29 May 2019, Sapporo, Japan

[26] S. Bravyi, O. Dial, J. M. Gambetta, D. Gil, Z. Nazario, "The Future of Quantum Computing with Superconducting Qubits Featured", Journal of Applied Physics, vol. 132, p. 160902, 2022

Chapter 10

Intelligent Quantum Information Processing

Future Directions of Research

Pampa Debnath, Arpan Deyasi,
and Siddhartha Bhattacharyya

10.1 CONCLUSION

As the name implies, a quantum computer primarily uses a number of quantum physical properties [1–4]. Given that they have processing speeds that are faster—even tenfold faster—than those of modern appropriate computers, they have the potential to be a very useful replacement. The phrase "quantum computing" is basically a synergistic blend of concepts from quantum physics, classical information theory, and computer science.

Information represented by qubits, the fundamental building block of a quantum computer, is processed through a process known as quantum information processing [4]. The core of information processing in the quantum realm is qubit encoding of the classical data, which derives the classical result from some quantum measurement operations and the intrinsic superposition and coherence features. Furthermore, the qubits' entanglement feature [5–7] facilitates long-distance communication at a faster data transfer rate. In the near future, distributed quantum networks will be implemented using this higher data transmission rate as the foundation. Furthermore, faster quantum communication is on the horizon since quantum information processing is being researched at a rapid pace, and quantum networks and internet services may one day be realized. Advances in single photon source and detector research have led to the proposal of novel algorithms on quantum key distributions that might facilitate the development of a reliable and secure communication system. Quantum cryptography, which is the foundation for creating quantum teleportation networks and aids in preventing eavesdropping, is derived from sequential single photon communication. While creating entangled photons at the chip level is still a difficult task, new optical fibers and semiconductor-based quantum dot detectors may be useful in the real-world application of encryption techniques [8–10].

In addition to device-level research, scientists have dedicated their efforts to induce intelligence in quantum information processing in order to make the systems resilient, fail-safe, and efficient [11–13]. Utilization of the essential properties of quantum computing in diverse machine learning

222   DOI: 10.1201/9781003373117-10

algorithms has been researched throughout the decades, consequently generating quantum-inspired/quantum computational intelligent algorithms. Recent research topics have included developing quantum neural networks, modeling quantum fuzzy principles, and evolving quantum metaheuristics [14–17]. With the development of these quantum algorithms, a new chapter in the history of intelligent information processing has begun, one in which the real-time performance of quantum information processing algorithms is improved by effectively fusing the concepts of quantum mechanics.

The methodological methods, theoretical investigations, mathematical and practical techniques, and applications of intelligent quantum information processing to engineering challenges are all brought together in this volume to showcase the latest developments and trends in the field. The book's focus is essentially on introducing various new hybrid quantum computational algorithms to overcome the shortcomings of traditional information processing algorithms. These include, but are not limited to, quantum networks, quantum machine learning, quantum information processing, quantum key distribution, quantum information processing, quantum encryption algorithms, and quantum knowledge discovery in databases. Through the use of illustrative examples and real-world case studies, it also seeks to highlight the advantages of the suggested ways above the most advanced currently available alternatives.

As earlier stated, quantum information theory is a broad study of the information processing capabilities of quantum systems derived from the characteristic quantum mechanical principles [18]. Present-day researchers make use of quantum superposition and interference to evolve quantum/quantum-behaved architectures, systems, and algorithms that are able to process information in a far better way compared to their classical counterparts. While almost every scientific and technological innovation encompassing the quantum computing paradigm resorts to the utilization of the principles of quantum superposition and interference, quantum entanglement seems to be the most trusted and investigated resource to date used for the purpose of secured data communication. Given the correlation between entangled states at distant points, it is possible to undertake simultaneous computing operations, which is otherwise impractical in any conventional system.

10.2 FUTURE RESEARCH INITIATIVES

The US government has been instrumental in supporting basic research in quantum information processing during the past two decades [19].

The focus has been invested in several challenges, including the development of mechanisms for quantum error correction, secured quantum communication, and the development of associated coherent quantum interfaces, in order to construct large-scale quantum computers and a quantum internet in the near future. In addition, scientists have embarked on the development of novel systems and materials for quantum information processing, thereby facilitating the means of precision measurement over the past few decades. Most importantly, the development of quantum simulators will be a much-sought affair as quantum computers grow in size and complexity. These simulators will have profound roles in simulating the components and subcircuits for the purpose of quantum characterization, validation, and verification, which is of paramount importance as far as the development of quantum technology is concerned [20].

Setting a goal for the next 20 years or so, efforts are being contemplated to design and develop relevant technology for large-scale, universal, fault-tolerant quantum computers capable of solving hard linear algebra problems, performing quantum simulations, and machine learning, which, otherwise, could not be performed classically [20, 21]. Such efforts would obviously give rise to large-scale, special-purpose quantum simulators, quantum annealers, and integrated quantum optical circuits for general-purpose quantum computers. In addition, theoretical advances would connect quantum information science with other fields including high-energy physics, quantum gravity, music, chemistry, and computational biology. Thus, the most challenging proposition for the coming 5–10 years will be the quantum construction, characterization, validation, and verification (qCVV) of highly coherent quantum annealers with tunable non-stochastic couplings [19].

As far as quantum communication is concerned, the focus should be targeted on envisaging continental length efficient and entangled quantum networks along with efficient coherent quantum interfaces to quantum computers [22]. The state-of-the-art quantum networks should be capable of serving as an efficient link between many quantum memories, high-speed quantum teleportation, cryptography, and modular quantum computing. On the contrary, smaller quantum networks would remain connected to the global quantum internet, which will address many other applications, including quantum digital signatures, quantum voting, secret sharing, anonymous transmission of classical information, and a host of sensing and metrology applications, in addition to providing reliable and secure communication and parallel computing.

Quantum sensing has established itself as an effective tool for brain and neuroscience, including real-time recording and imaging of action potentials. Quantum sensing finds use in a number of applications, including life sciences, chemistry, and batteries, to name a few. Research and development of quantum sensing and metrology are also on the cards.

To be precise, there is a need for coordination of theoretical investigations and advances with sound experimental and technological efforts in the domains of developing efficient quantum error correction codes, methods for quantum communication, quantum sensing, and imaging, in order to attain the final grand challenges discussed previously.

REFERENCES

[1] Z. Meng, "Review of Quantum Computing", 13th IEEE International Conference on Intelligent Computation Technology and Automation, 2020, Xi'an, China

[2] R. Rietsche, C. Dremel, S. Bosch, L. Steinacker, M. Meckel, J.-M. Leimeister, "Quantum Computing", Electronic Markets, vol. 22, pp. 2525–2536, 2022

[3] A. Erhard, J. J. Wallman, L. Postler, M. Meth, R. Stricker, E. A. Martinez, P. Schindler, T. Monz, J. Emerson, R. Blatt, "Characterizing Large-Scale Quantum Computers via Cycle Benchmarking", Nature Communications, vol. 10, no. 5347, 2019

[4] A. K. Pati, S. Braunstein, "Role of Entanglement in Quantum Computation", Journal of the Indian Institute of Science, vol. 89, no. 3, pp. 295–302, 2009

[5] B. Wong, "On Quantum Entanglement", International Journal of Automatic Control System, vol. 5, no. 2, pp. 1–7, 2019

[6] N. Zou, "Quantum Entanglement and Its Application in Quantum Communication", Journal of Physics: Conference Series, vol. 1827, p. 012120, 2021

[7] L. Hadjiivanov, I. Todorov, "Quantum Entanglement", Bulgarian Journal of Physics, vol. 42, pp. 128–142, 2015

[8] N. Gisin, R. T. Thew, "Quantum Communication Technology", Electronics Letters, vol. 46(14), pp. 965–967, 2010

[9] N. Gisin, R. T. Thew, "Quantum Communication Technology", Electronics Letters, vol. 46(14), pp. 965–967, 2010

[10] J. Chen, "Review on Quantum Communication and Quantum Computation", Journal of Physics: Conference Series, vol. 1865, no. 022008, 2021

[11] A. Dey, S. Bhattacharyya, S. Dey, J. Platos, V. Snasel, "A Quantum Inspired Differential Evolution Algorithm for Automatic Clustering of Real Life Datasets", Multimedia Tools and Applications, pp. 1–30, 2023, https://doi.org/10.1007/s11042-023-15704-3

[12] A. Dey, S. Bhattacharyya, S. Dey, D. Konar, J. Platos, V. Snasel, L. Mrsic, P. Pal, "A Review of Quantum-Inspired Metaheuristic Algorithms for Automatic Clustering", Mathematics, vol. 11, p. 2018, 2023, https://doi.org/10.3390/math11092018

[13] T. Dutta, S. Bhattacharyya, B. K. Panigrahi, I. Zelinka, L. Mrsic, "Multi-Level Quantum Inspired Metaheuristics for Automatic Clustering of Hyperspectral Images", Quantum Machine Intelligence, vol. 5, p. 22, 2023, https://doi.org/10.1007/s42484-023-00110-7

[14] P. Pal, S. Bhattacharyya, J. Platos, V. Snasel, "A Brief Survey on Image Segmentation Based on Quantum Inspired Neural Network", International Journal of Hybrid Intelligence, vol. 2, no. 2, pp. 102–118, 2023, https://doi.org/10.1504/IJHI.2023.129296

[15] D. Konar, A. Das Sarma, S. Bhandary, S. Bhattacharyya, A. Cangi, V. Aggarwal, "A Shallow Hybrid Classical-Quantum Spiking Feedforward Neural Network for Noise-Robust Image Classification", Applied Soft Computing, vol. 136, 2023, https://doi.org/10.1016/j.asoc.2023.110099

[16] D. Konar, S. Bhattacharyya, T. Gandhi, B. K. Panigrahi, R. Jiang, "3D Quantum Inspired Self-Supervised Tensor Network for Volumetric Segmentation of Medical Images", IEEE Transactions on Neural Networks and Learning Systems/*TechRxiv*, January 24, 2023, https://doi.org/10.1109/tnnls.2023.3240238

[17] D. Konar, S. Bhattacharyya, B. K. Panigrahi, E. C. Behrman, "Qutrit-Inspired Fully Self-Supervised Shallow Quantum Learning Network for Brain Tumor

Segmentation", IEEE Transactions on Neural Networks and Learning Systems, vol. 33, no. 11, pp. 6331–6345, 2022, https://doi.org/10.1109/TNNLS.2021.3077188

[18] I. B. Djordjevic, "Quantum Information Processing Fundamentals", Quantum Information Processing, Quantum Computing, and Quantum Error Correction, chapter 3, pp. 125–158, 2021

[19] https://basicresearch.defense.gov/Portals/61/Documents/future-directions/Future_ Directions_Quantum.pdf?ver=2017-09-20-003031-450

[20] P. W. Shor, "Polynomial-Time Algorithms for Prime Factorization and Discrete Logarithms on a Quantum Computer", SIAM Review, vol. 41, no. 2, pp. 303–332, 1999

[21] F. E. Magniez, M. Santha, M. Szegedy, "Quantum Algorithms for the Triangle Problem", SIAM Journal of Computing: C, vol. 37, no. 2, pp. 413–424, 2007

[22] S. Wehner, D. Elkouss, R. Hanson, "Quantum Internet: A Vision for the Road Ahead", Science, vol. 362, no. 6412, 2018

Index